Extending the Frontiers of Mathematics

Inquiries into proof and argumentation

Extending the Frontiers of Mathematics

Inquiries into proof and argumentation

Edward B. Burger

Williams College

Key College Publishing
Innovators in Higher Education

www.keycollege.com

in cooperation with

Springer

Edward B. Burger
Department of Mathematics
Williams College
Williamstown, MA 01267

Key College Publishing was founded in 1999 as a division of Key Curriculum Press® in cooperation with Springer New York, LLC. We publish innovative texts and courseware for the undergraduate curriculum in mathematics and statistics as well as mathematics and statistics education.
For more information, visit us at www.keycollege.com.

Key College Publishing
1150 65th Street
Emeryville, CA 94608
(510) 595-7000
info@keycollege.com
www.keycollege.com

Published by Key College Publishing, an imprint of Key Curriculum Press.

Development Editors: Allyndreth Cassidy, Kristin Burke
Editorial Production Project Manager: Laura Ryan
Text Design, Composition, and Art: Happenstance Type-O-Rama
Cover Designer: Jensen Barnes
Cover Photo Credit: Jensen Barnes/Edward B. Burger
Copyeditor: Tara Joffe
Indexer: Victoria Baker
Printer: Data Reproductions Corporation
Editorial Director: Richard J. Bonacci
General Manager: Mike Simpson
Publisher: Steven Rasmussen
ISBN: 1-59757-042-7

Library of Congress Cataloging-in-Publication Data
Burger, Edward B., 1963–
Extending the frontiers of mathematics : inquiries into proof and argumentation /
Edward B. Burger.
p. cm.
Includes index.
ISBN-13: 978-1-59757-042-8
ISBN-10: 1-59757-042-7
1. Proof theory. 2. Mathematical analysis—Foundations. 3. Logic, Symbolic and mathematical. I. Title.
QA9.54.B87 2007
511.3'6—dc22

2006045729

Printed in the United States of America
10 9 8 7 6 5 4 3 2 1 11 10 09 08 07 06

Contents

Introduction

A brief travel guide to the journey ahead

More than half a century ago, Professor R.L. Moore from The University of Texas at Austin pioneered a style of inquiry-based learning that developed critical thinking skills and creativity. Moore's method involved a delicate balance between the Socratic method and an independent exploration in which students took an active role in their education by discovering mathematical ideas for themselves, without any lectures or texts. Extending Moore's original educational template, this small book invites its readers to experience the process involved in pursuing mathematical research and to develop thinking strategies that can be employed to generate new mathematical ideas. In the section "To the instructor," instructors will find a description of a variety of courses for which this book can be used.

In the modules ahead, we will be challenged to bring ourselves to discover some beautiful mathematical vistas. The journey will open with conundrums that at once entertain and offer the opportunity to be creative within a mathematical context. Next we will be introduced to the basic notions of mathematical formalism, and from there, we will be ready to explore a world of wondrous mathematical ideas.

This introduction offers a panoramic overview of what conceptually lies ahead. The commentary that follows contains notions that might be foreign to some readers upon first reading. These ideas are carefully defined and explained in Module 2. The goal of this "putting the cart before defining the horse" writing style is to encourage students to read the following passage twice: once now and then again after exploring Module 2.

To the student: A great adventure

This book attempts to reflect a natural progression of mathematical thought. It is overflowing with statements—mathematical declarative sentences that are either true or false. In the real (but often abstract) world of research mathematics, we do not know, in advance, which statements are true and which are false. Thus when we are in the happy—albeit artificial—homework scenario in which someone asks us to "prove a theorem," we have an enormous advantage: We are told that the statement is true! Half of our work has been done for us, and the *only* task that remains is to devise a proof. Of course the word *only*, in most cases, is utterly inappropriate, because crafting a mathematical proof is an extremely challenging and creative endeavor.

Mathematicians on the front lines of mathematical discovery do not have the luxury of knowing *a priori* if their assertions are true or false. One of the most important steps in research is to conjecture what is the truth and then attempt to verify it by hunting down a proof. This book offers its readers the experience of facing statements and determining whether those statements are true or false. The preamble to each statement to come is, "Prove and extend *or* disprove and salvage." If we believe a statement is true, then the challenge at hand is to produce a mathematical proof of its validity. If we believe a statement is false, then the mission is to exhibit an explicit counterexample. In either case, there is an important follow-up challenge that is designed to instill good mathematical habits: If the statement is true, can we extend it? If the statement is false, can we salvage it?

Extending or generalizing a theorem involves producing a new result that has within it the original theorem as a special case. To generalize a result, we might consider weakening the hypotheses or wonder if the converse also holds (thus producing a biconditional—"if and and only if"—result). Alternatively, we might consider strengthening the conclusion: If a result states that there exists a *blippity-blop*, can we show how to explicitly generate that *blippity-blop* or can we show that the *blippity-blop* is unique or can we give an upper bound for how many such *blippity-blops* there are? This type of thinking often leads to new and interesting extensions.

Salvaging a statement is, in some sense, the other side of the logical coin. Once we have a counterexample in hand, we can study it and wonder if there are any additional (ideally, mild) conditions we can add to the hypotheses that would allow us to establish the desired conclusion. Alternatively, perhaps the stated conclusion is too strong—is there a mild weakening of the conclusion that can be shown to hold? If a statement contains the phrase "if and only if" and one direction can be shown to be false, then certainly just establishing the one conditional that holds is fine; however, even better is to produce a modified biconditional "if and only if" statement that is true. A statement is not truly salvaged until a proof of the modified statement is found. Thus for each and every statement, we are to associate a proof (either a proof of the original statement or a proof of our salvaged version).

The journey ahead is challenging and will lead to moments of great intellectual triumphs, as well as moments of deep frustration. Both emotions are natural in this creative venture. The discovery approach provides a rich and deep understanding, but it does so through *uncomfortable learning*. First, there is very little explicitly explained throughout these pages. Second, much of this material, indeed most of mathematics, is genuinely difficult. Without footholds and the comfort of a conventional text, some challenges may seem daunting. Failure is crucial: Create and then try an idea, fail, learn from that misstep, and try again. Learn from failure—it truly is the best teacher. One of the most powerful ways to develop insight and intuition is through a careful examination of previous failed attempts.

There are, however, some narrow footholds to help readers along the way. A few hints are given within the body of the text, but many more (together with leading questions and commentary) are offered at the end of the book. Those challenges that have some additional material included at the close of the book are not marked with any special symbol. This way, readers, *if they wish*, can consider each challenge without any distractions, and only if they become completely stuck can they check the back of the book to see if a helpful comment is waiting. Of course the hints, leading questions, and commentary at the end are there to be used whenever and however readers desire. Perhaps even more helpful than the end-of-the-book hints are the previous challenges themselves. The challenges unfold in a very particular order and build on each other. Therefore when readers do not know how to move forward, they may often find it helpful to consider where they just were.

Our hope is that these challenges will foster interesting discussions—whether they be between student and instructor or student and student. Part of our main goal is to invite readers to articulate their ideas in a clear fashion. The fundamental questions to ponder when faced with a mathematical proof (one's own or another's) are: *Is it correct? Is it complete? Is it clear?* These questions are the true test of an argument. Clarity is an important element in moving the borders of mathematics outward—we need to be able to communicate our ideas to others. Appendix 2 offers a proof primer with detailed commentary and suggestions on how to create, write, and then polish original mathematical arguments.

In the best of all possible worlds, readers will come away from their experience using this book with a deeper understanding of mathematics, a curiosity to explore more advanced mathematical ideas, and a greater passion and talent for imaginative thinking and for articulating their creative ideas so as to inspire others.

To the instructor: The mathematics to come

This book can be employed in a variety of ways. It can be the text for a course taught using a discovery approach, in which students present their results during class meetings. It can also be used as a supplemental text for courses that either are lecture-based or have a combination of lectures and student presentations. Finally, the book can be used for independent study. In whatever context this text is used, the following features can develop and encourage mathematical thinking and curiosity:

- Puzzles and conundrums open the text, honing the student's ability to create original mathematical arguments.
- "Prove and extend *or* disprove and salvage" instruction is a recurring theme throughout, providing a consistent framework for approaching mathematics.
- Challenges of various levels of difficulty foster mathematical discovery and lively discussion.

Appendix 3 includes specific details for instructors, describing how this material has been used by the author in the classroom. In this appendix, instructors will find specific suggestions for how this text can be incorporated into a variety of different teaching styles. In addition, detailed pointers are included to help enhance the classroom experience if a pure discovery approach is being employed. Additional instructor resources are available through the publisher.

The mathematics introduced in these pages cover a wide range of areas. Certainly no course can cover all the material in one semester. If adopting a pure discovery approach, then one can reasonably expect to cover roughly 10 or 11 modules (depending on the individual class, the class size, and the goals of the course). To help get the curricular juices flowing, the following list offers several different course templates and suggested modules that might be particularly appropriate, depending on the emphasis of the course. There are also optional modules that an instructor might elect to use—either in their entirety or just selected portions of the material—or to omit completely.

Introduction to Mathematical Proof

Modules 1, 2, 3, 4, 6, 7

Optional Modules 5, 8, 11, 13, 14, 15

Introduction to Discrete Mathematics

Modules 1, 2, 3, 4, 6, 7, 10, 11

Optional Modules 8, 12, 15

Survey of Mathematical Foundations

Modules 1, 3, 4, 5, 6, 7, 11, 14, 16, 18

Topics in Mathematics for Future Teachers

Modules 1, 2, 4, 6, 7, 10, 14, 16, 18

Optional Modules 3, 15

Capstone Senior Seminar in Mathematics

Modules 1, 6, 7, 8, 14, 16, 17, 18, 19

Optional Module 15

No matter in what context this book is used, I hope it will invite readers to enhance and hone their abilities to be creative and inspire them to strive to move the frontiers of mathematics forward. Enjoy the mathematical adventure!

—Edward Burger

1

Puzzles and patterns

A precursor to proofs

There are two fundamental components involved in generating mathematics: The first is the creative thinking process, which allows us to develop original ideas and insights and, in turn, to move the frontiers of our knowledge forward. The second component is the formalism and logical rigor required in crafting mathematical arguments and producing proofs. The journey through this text offers readers an opportunity to develop and refine these two important skills—creating new ideas and composing clear proofs. In Module 2, we will build the basic formalism and logic necessary to prove mathematical theorems. In this module, we celebrate the wild creativity and imagination involved in making breakthroughs. Your mission, if you decide to accept it, is to face the following ten challenges and do your best to be as creative as possible.

Some of the challenges are indeed challenging. Don't become overly frustrated—keep the degree of frustration at a healthy level. Experiment, consider examples, look at simpler scenarios, devise thinking strategies that might allow you to gain some insight into the challenge at hand. Don't be afraid to fail! The best ideas and insights often arise out of failed attempts and false starts. Fail agressively! Your primary goal is not to resolve each and every challenge (although if you do, congratulations!). Rather the goal is to generate as many approaches as you can and refine your abilities to look at issues from as many vantage points as possible. The main theme in this opening module is: What do you do when you don't know what to do? Think, fail, create, and enjoy!

Senators and snakes

1.1. An honest day's work in the senate. On one particularly sober day in the senate chamber, all 100 senators were seated and uncharacteristically silent. The following two statements had just been released by a nonpartisan government watchdog organization:

- There exists at least one honest senator.
- Given any two senators, at least one of them is dishonest.

In fact, each of the two previous statements is true. In view of these statements, your challenge is to see if you can determine exactly how many of the 100 senators are honest and how many are dishonest. If those numbers can be determined using only the given information, then find the number of honest senators. If the number cannot be found, explain why not. Do your answers help explain why the senators sat in silence?

1.2. Indianapolis Jones and the box of snakes. As everyone well knows, Indianapolis Jones, the archaeologist and obsessive-compulsive treasure hunter, hates snakes. After a long and exciting quest to uncover just one of the pieces of the priceless and ancient *Vass* vase, Indianapolis Jones found himself face-to-face with two ornate boxes (so, more accurately, he was face-to-boxes). One box was embossed with a coat of arms that contained an image of a diamond, while the other was marked with an image of two crossed swords. Given all of his research, he knew that each box contained either a precious *Vass* piece or a family of 100 deadly snakes that instantly kills anyone whose nickname is a city or metropolitan area. Indianapolis had no choice—he had to open one box. He reached into his pocket and removed the final clue—a piece of parchment with an ominous message written in Sanskrit. Indianapolis read two statements, which, when translated into English, were, "At least one of the boxes contains a treasured *Vass* piece. But beware: A colony of poisonous snakes awaits you in the diamond-embossed box." Indianapolis had deduced from previous clues that either both statements are true or both are false. Your challenge is to determine which box

Indianapolis Jones should open in order for him to *safely* accomplish his ultimate goal: To get a *Vass* vase.

Checkerboards undercover

1.3. Covering cut boards. Suppose we are given a standard 8×8 checkerboard and an immense supply of dominoes. Each domino can cover exactly two adjacent squares on the checkerboard.

One domino covers two adjacent squares.

As a warm-up, verify that the checkerboard can be covered completely by dominoes, with each square covered by exactly one domino and each domino resting on exactly two squares. Assume next that two squares of the checkerboard have been cut off, as shown.

Your challenge now is to determine whether you can cover this cut checkerboard with dominoes so again each square is covered by exactly one domino and each domino is resting on exactly two squares. Finally, your last challenge is to consider the same question for the following truncated checkerboard.

Does your answer change? Justify your answers.

1.4. An average square. Imagine a checkerboard that is infinitely expansive—it extends forever in every direction.

A natural number (1, 2, 3, 4, ...) is written in each square. Notice that each square is surrounded by four squares that share one of the first square's sides.

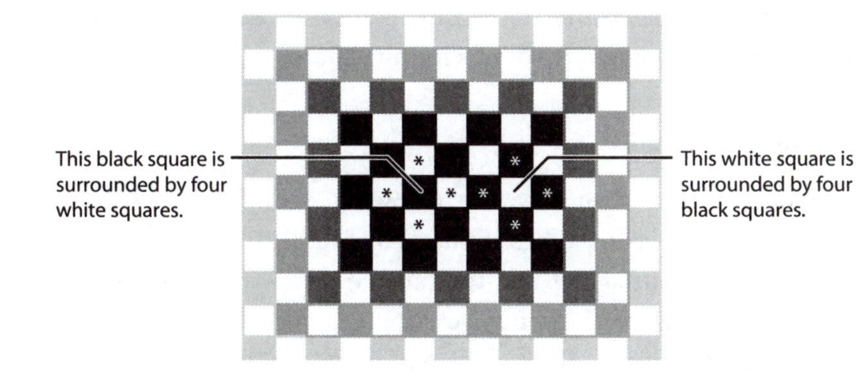

Each white square is surrounded by four black squares, and each black square has four white squares around it. The natural numbers written in the squares are arranged on this endless checkerboard so that the number written in a square is precisely the average value of the numbers written in the four surrounding squares. So, for example, if a black square were to be surrounded by white squares labeled 7, 4, 8, and 5, then the number written in that black square would be 6 (the average of 7, 4, 8, and 5).

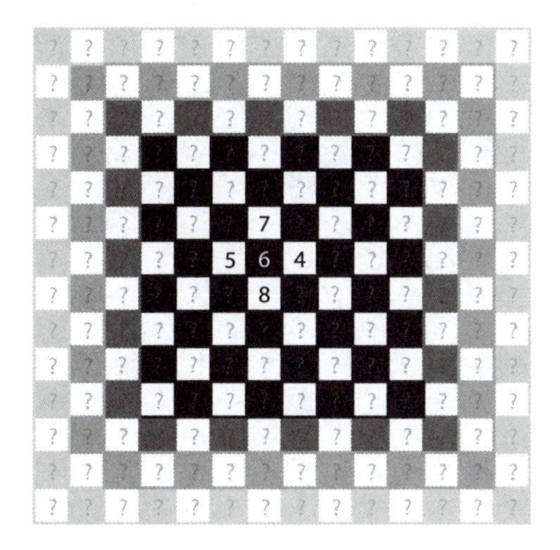

Your challenge is to determine the most you can conclude about the natural numbers written in the squares of this infinite board.

Flipping cards and coins

1.5. Magic revealed. Consider the following mathematical illusion: A regular deck of 52 playing cards is shuffled several times by an audience member until everyone agrees that the cards are completely shuffled. Then, without looking at the cards themselves, the magician divides the deck into two equal piles of 26 cards. The magician taps both piles of face-down cards three times. Then, one by one, the magician reveals the cards of both piles. Magically, the magician is able to have the cards arrange themselves so that the number of cards showing black suits in the first pile is identical to the number of cards showing red suits in the second pile. Your challenge is to figure out the secret to this illusion and then perform it for your friends.

1.6. Heads up with a blindfold. Some number of coins are spread out on a table. They lie either heads up or tails up.

Unfortunately you are blindfolded, and thus both the coins and the table upon which they sit are hidden from view. Certainly you can feel your way across the table and count the total number of coins on the table's surface, but you cannot determine if any individual coin rests heads up or down (perhaps you're wearing gloves). You are informed of one fact (beyond the total number of coins on the table): Someone tells you the number of coins that are lying heads up. You can now rearrange the coins, turn any of them over, and move them in any way you wish, as long as the final configuration has all the coins resting (heads or tails up) on the table. Your challenge is to turn over whatever coins you wish and divide the coins into two collections so that one collection of coins contains the same number of heads up coins as the other collection contains.

Juggling balls

1.7. Weighing your options. You find yourself on a reality TV show that has you competing with other real people in totally artificial circumstances. In one scenario, you are given nine balls of clay. You are informed by the program's B-celebrity host that hidden inside one of those clay balls is a key that will unlock a refrigerator that houses a vast quantity of food. Since the producers "thought" the ratings would be higher if the contestants were deprived of nutrition, even the thought of brussels sprouts makes your mouth water. You are told that the eight balls that do not contain the key to your dietary dreams all weigh the same. The special ball with the key inside weighs slightly more, but not enough for you to feel the difference by holding the balls in your hand. One of the program's sponsors, *Replace-Oh!*, the company that manufactures one-time-use balance scales (with the slogan, "Weigh aweigh then throw away!"), has agreed to provide some of its scales in exchange for a few shameless plugs throughout the program. Their scales will tell which side is heavier and then instantly self-destruct.

Scale before use

Scale after one use

You are only allowed to break open one clay ball to see if you can find the refrigerator key. Your challenge is to determine the fewest disposable balance scales required to guarantee that you can identify the ball with the key. Justify your answer.

1.8. Being on the ball. You are given infinitely many Ping-Pong balls, each labeled with a natural number and lined up in order: 1, 2, 3, … . You are also provided with a very large barrel. You are now asked to imagine a sequence of events that cannot physically transpire. Imagine that in 1 minute, you will perform infinitely many tasks. The first task, which you have 30 seconds to complete, requires you to drop the first 10 Ping-Pong balls into the barrel (those balls numbered 1 through 10) and then immediately reach in and discard the ball numbered 1. In the next 15 seconds, you are to drop in the next 10 balls (numbered 11 through 20) and then quickly reach into the barrel and remove ball number 2. In half the time that remains (7.5 seconds), you are to pour in the next 10 balls (21 through 30) and instantly reach in and remove ball 3. This process continues, with the amount of time allotted for a particular task being equal to half the time allowed for the previous task. Once the minute has transpired, your work is done. Now that you're done, your challenge is to imagine looking into the barrel and determining how many Ping-Pong balls are left. *Super Bonus:* Once you've addressed this challenge, consider the following modification: The balls are not numbered (the balls are indistinguishable from one another) and at each stage you reach in and remove *any* ball. Does your answer change?

Counting on a good hand and a good grade

1.9. Shaking down their friends. Carol and Chris are roommates who share a passion for Texas Hold 'Em poker. One evening, they organize a small poker tournament in their dorm and invite eight friends—four pairs of roommates. When everyone arrives and before they shuffle up and deal, there's lots of handshaking. Although we have no idea who shook hands with whom, we do know that no one shook hands with him- or herself, and no one shook hands with his or her own roommate. Given these facts, someone might shake as few as no hands or as many as eight other people's hands. At midnight, Chris gathered the crowd and asked the nine other poker players how many hands each of them had shaken. Much to Chris's amazement, each person gave a different answer. That is, someone didn't shake any hands, someone else shook one hand, someone else shook two hands, someone else shook three hands, and so forth, down to the last person, who shook eight hands. Given this outcome, your challenge is to determine the exact number of hands that Carol, Chris's roommate, shook.

1.10. Grading off the top of your head. A cruel calculus instructor decided to terrorize her students. The instructor announced that during the next class, the students will line up, facing away from the front of the line.

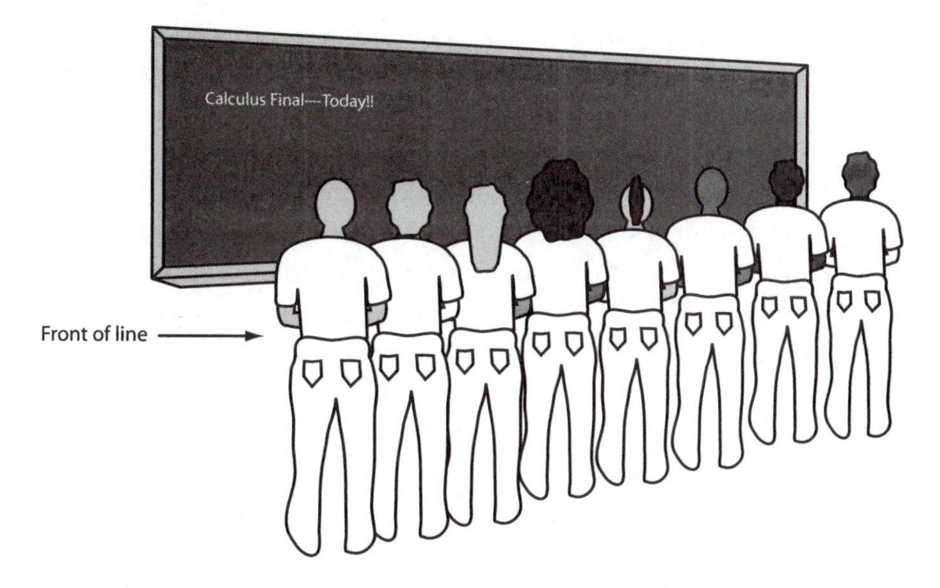

The instructor will then place either a white or a gray dunce cap on each student's head. Each student will be unable to see his or her own cap but will be able see the cap colors of all those classmates who are in front of him or her.

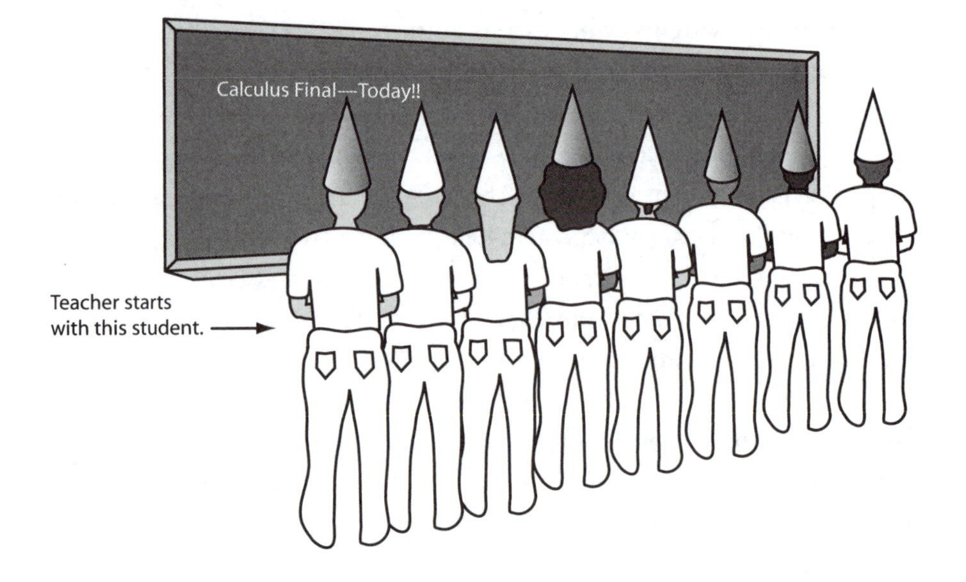

Starting at the head of the line, each student will be asked, in turn, "What is the color of your dunce cap?" Students will only be allowed to respond by saying "white" or "gray." The students who answer correctly will be given A's, and the students who answer incorrectly will fail the course. For each student, the instructor will respond with either "Correct! You receive an A" or "You fail! Get out of my classroom, you dunce," depending on the correctness of the answer. All students will be able to hear all students' answers and the instructor's responses. Knowing this horrific fate that awaits them, the students have all night to come up with a plan. If there are n students in the class, how many of them can be guaranteed to receive A's? Your challenge is to devise a scheme that the students can employ to allow as many of them as possible to receive A's. (*Caution:* No "cheating" is allowed; that is, students cannot use the tones of their voices or say additional phrases or use hand gestures to provide any additional information.)

Stepping back

List all the thinking strategies you employed to build insight into the challenges and (we hope) to resolve them. Our mathematical creativity will flow from our ever-evolving ability to generate new ideas and uncover hidden structure and patterns.

2

Bringing theorems to justice

Exposing the truth through rigorous proof

Establishing the validity of statements is at the heart of all mathematics. Here we introduce and develop the basic techniques and tools of the trade that can be applied in the artful act of creating a mathematical proof. Even the seemingly simple-sounding acts of understanding what constitutes a "mathematical statement" and parsing such statements are serious challenges. We begin at the beginning and introduce the basic language of mathematics that we will speak.

Making a statement

In mathematics, we define a *statement* as "a declarative sentence that is either true or false but not both."

2.1. Determine which of the following sentences are mathematical statements:

- My e-mail password is "swordfish."
- I don't understand.
- Are you really taking that math course? Are you crazy?
- This sentence is false.
- $310 \div 0 = 31$

- I prefer pi.
- If you become a mathematics major, then you'll land a high-paying job.

Bonus Puzzler: Which sentence stands out as "wow"?

If A is a statement, then the opposite statement is called its *negation* and is defined to be the statement "Not A." For example, if A = "The natural number n is odd," then its negation is the statement "Not (the natural number n is odd)"; or, more grammatically correct, "It is not the case that the natural number n is odd"; or simply, "The natural number n is not odd." Notice that it might be tempting to assert, "The natural number n is even." However, that is *not* the logical negation of the original statement. In fact, that statement is a *theorem*—albeit a small one—that every natural number is either odd or even.

2.2. Produce as many grammatically correct versions as you can for the negation of each statement:

- The number x is negative.
- For all x, $f(x) < 0$.
- There exists an x such that $g'(x)$ is undefined.

A *compound statement* is a mathematical statement that contains several mathematical statements. For example, if A and B are two statements, then we can combine them to create the compound statement "A and B" or the compound statement "A or B." The most important compound statement is the *implication statement:* "If A, then B." In this implication, statement A is called the *hypothesis,* and statement B is called the *conclusion.* For example, if A = "$f(x) = x^2 + 1$" and B = "$f(3) = 10$," then the statement "If A, then B" is the statement "If $f(x) = x^2 + 1$, then $f(3) = 10$."

Given the implication "If A, then B," we define the *converse* to be the statement "If B, then A"; the *inverse* to be the statement "If not A, then not B"; and the *contrapositive* to be the statement "If not B, then not A." So given the statement "If $f(x) = x^2 + 1$, then $f(3) = 10$," its converse is the statement "If $f(3) = 10$, then $f(x) = x^2 + 1$"; its inverse is the statement "If $f(x) \neq x^2 + 1$, then $f(3) \neq 10$"; and its contrapositive is the statement "If $f(3) \neq 10$, then $f(x) \neq x^2 + 1$."

(*Editorial comment:* After this module, I predict that in your mathematical work, you will never use the word "inverse" in this context.)

The compound statement "*A* if and only if *B*" is defined to be the statement "(If *A*, then *B*) *and* (if *B*, then *A*)." We say that the statements *A* and *B* are *equivalent* if the statement "*A* if and only if *B*" is true.

The truth, the whole truth, and other stuff besides the truth

Recall that a mathematical statement is either true or false, but not both. Given a statement *A*, the statement "Not *A*" is true whenever *A* is false and is false whenever *A* is true. Thus if the statement *A* is true, then the statement "Not *A*" is false; if *A* is false, then "Not *A*" is true. We can record these various cases in terms of a table (known as a *truth table*), in which all possibilities are listed.

A	Not *A*
T	F
F	T

Truth table for Not *A*

If *B* is a statement, then the statement "*A* and *B*" is true if *both* *A* and *B* are true and is false otherwise. In this case, there are four possible scenarios for a truth table: *A* could be true, and *B* could be true; *A* could be true, and *B* could be false; *A* could be false, and *B* could be true; *A* could be false, and *B* could be false.

A	*B*
T	T
T	F
F	T
F	F

All four possible truth values for the statements *A*, *B*

Here is the truth table for "*A* and *B*" in these four cases:

A	*B*	*A* and *B*
T	T	T
T	F	F
F	T	F
F	F	F

Truth table for "*A* and *B*"

The statement "*A* or *B*" is true if *either A* or *B* is true and is false if *both A* and *B* are false.

A	*B*	*A* or *B*
T	T	T
T	F	T
F	T	T
F	F	F

Truth table for "*A* or *B*"

Two statements are equivalent if their corresponding truth tables are identical.

2.3. Express the statement "Not (*A* and *B*)" as an equivalent statement involving only "not" and "or." Then express the statement "(*A* or *B*)" as an equivalent statement involving only "not" and "and." Justify your answers with truth tables.

The implication "If *A*, then *B*" is false whenever the hypothesis *A* is true while the conclusion *B* is false. In all other cases, the implication is true.

A	B	If A, then B
T	T	T
T	F	F
F	T	T
F	F	T

Truth table for "If A, then B"

2.4. Which of the following implications are true?

- If $1 + 1 = 3$, then $\sqrt{36} = -6$.
- If $\cos \pi = 0$, then $\sin \pi = 0$.
- If $\sin \pi = 0$, then $\cos \pi = 0$.
- If $x = -3$ satisfies $x^2 = 9$, then $\log_2 \frac{1}{8} = -3$.

2.5. Consider an implication, its converse, its inverse, and its contrapositive. Which of these statements are equivalent to one another? Justify your answers with truth tables.

Our fundamental quest in mathematics is to consider interesting statements—either our own statements or statements of others—and then determine whether those statements are true. If a statement is true, then once we have established its truth, we try to extend or generalize it to produce a deeper result. If a statement is false, then once we have exhibited a counterexample, we try to salvage it by either including additional hypotheses or weakening the conclusion. Thus at the heart of mathematics, we find *proofs*—justifications that establish the truth of mathematical statements. Most mathematical statements can be expressed as implications: If (hypotheses), then (conclusion). Once we verify a statement to be true, we refer to it as a *theorem*. (Actually, theorems come in many shapes and sizes. A tiny theorem that is used to prove a larger result is often known as a *lemma*. A nice consequence that follows quickly from a theorem is called a *corollary*. Still other theorems are referred to as *propositions*, but don't ask me why.)

Arguing the case through proof

There are many different methods we can employ to prove an implication. A *direct proof* is one in which we directly apply the hypotheses and deduce the truth of the conclusion. A *proof by contradiction* is one in which we assume that the negation of the conclusion is true and use that assumption, together with the hypotheses, to deduce the truth of a statement previously known to be false—thus leading to a contradiction. The impossible situation of having a statement that is at once true and false implies that our assumption was false, and hence the desired conclusion must be true.

To illustrate these proof techniques and develop our ability to prove theorems, we consider several results regarding familiar objects. We define the set of *natural numbers,* denoted $\mathbb{N}$, to be the collection defined by $\mathbb{N} = \{1, 2, 3, \ldots\}$ and the set of *integers,* denoted $\mathbb{Z}$, to be the collection defined by $\mathbb{Z} = \{\ldots, -3, -2, -1, 0, 1, 2, 3, \ldots\}$. The *perfect squares* are the integers 0, 1, 4, 9, 16, 25, $\ldots, n^2, \ldots$.

STATEMENT. *If n is an integer, then* $2^n \leq 3^n$.

Commentary: On first blush, this statement sounds trivially true. However, careful consideration will reveal otherwise. If we consider $n = -1$, then the statement would assert that $\frac{1}{2} \leq \frac{1}{3}$, which is clearly false. How can we salvage the statement into a theorem? There are many ways; here are two:

THEOREM. *If n is a positive integer, then* $2^n \leq 3^n$.

Notice that in this theorem, we strengthened the hypothesis. The version we will prove is:

THEOREM. *If n is an integer satisfying $n \geq 0$, then* $2^n \leq 3^n$.

Proof. We begin by observing that the result holds in the case $n = 0$: $2^0 = 3^0 = 1$. Thus we need only consider the case in which $n > 0$. In this case, we know that

$$2^n = \underbrace{2 \times 2 \times 2 \times \cdots \times 2}_{n \text{ factors}}$$

Because we know that $2 < 3$, we have

$$\underbrace{2\times 2\times 2\times\cdots\times 2}_{n\text{ factors}} < \underbrace{3\times 3\times 3\times\cdots\times 3}_{n\text{ factors}},$$

and hence $2^n < 3^n$. Therefore for all $n \geq 0$, we have $2^n \leq 3^n$, which completes our proof.

The previous proof is an example of a direct proof—we used what we knew directly to deduce the conclusion we desired. Our next illustrative statement is:

STATEMENT. *If n is a positive integer, then the smallest perfect square that exceeds n^2 is $n^2 + 2n + 1$.*

Commentary: If we try some examples, we will see that the result seems to hold true. If $n = 1$, then $n^2 = 1$ and $n^2 + 2n + 1 = 4$; if $n = 2$, then $n^2 = 4$ and $n^2 + 2n + 1 = 9$; if $n = 3$, then $n^2 = 9$ and $n^2 + 2n + 1 = 16$. Although these examples do not prove the result in general, they do allow us to build some insight. We now prove:

THEOREM. *If n is a positive integer, then the smallest perfect square that exceeds n^2 is $n^2 + 2n + 1$.*

Proof. Given a positive integer n, we notice that by basic algebra we have that $(n + 1)^2 = n^2 + 2n + 1$. Thus we conclude that $n^2 + 2n + 1$ is indeed a perfect square. Moreover, because $n > 0$, we also have that $n^2 < n^2 + 2n + 1$. Therefore we need now only establish that $n^2 + 2n + 1$ is the *smallest* perfect square exceeding n^2. We proceed by contradiction. Assume there exists a positive integer m such that

$$n^2 < m^2 < (n+1)^2.$$

Recalling from calculus that the square root function is positive and increasing, if we take square roots of the previous inequality, we have that $n < m < n + 1$. Thus we find an integer m strictly between n and $n + 1$. However, n and $n + 1$ are consecutive integers, and therefore there is no integer between them. Thus we have a contradiction, which implies that our assumption must be false. Hence there does *not* exist a positive integer m such that $n^2 < m^2 < (n + 1)^2$. That is, $(n + 1)^2$ is the *smallest* perfect square that exceeds n^2, which completes our proof.

Now that we see the result is true, we can attempt to extend it. There are many extensions. Here is one such generalization:

THEOREM. *Let k be a positive integer. If n is a positive integer, then the kth perfect square that exceeds n^2 is $n^2 + 2kn + k^2$.*

Observe how the previous proofs were written in complete sentences that logically flow together. Although creating a proof is truly an art form and there are many styles, it is crucial that the arguments are offered in a clear, well-written manner. Appendix 2 offers some additional commentary on the basics of writing and thinking like a mathematician.

Often the best way to get started in this proof business is to first consider various cases and then construct examples to help build intuition and insights into why a statement might be true in general. Occasionally, through examples, a general pattern emerges—always attempt to uncover patterns. Now it's your turn! We say an integer is *even* if it is a multiple of 2 (that is, 2 multiplied by some integer), and otherwise, it is *odd*.

2.6. Prove and extend *or* disprove and salvage:

STATEMENT. *The sum of two odd numbers is an even number.*

2.7. Prove and extend *or* disprove and salvage:

STATEMENT. *The product of any three consecutive integers is a multiple of* 3.

2.8. Prove and extend *or* disprove and salvage:

STATEMENT. *If the average of four distinct integers is* 94, *then at least one of the integers must be greater than or equal to* 97.

In mathematics, statements that appear trivial or self-evident are often the most challenging to establish with a rigorous proof. To illustrate this assertion, we recall that the set of integers $\mathbb{Z}$, together with the operations of addition and multiplication, satisfies a number of very important algebraic properties. In particular, with respect to addition, we have the following:

- Addition is *associative*; that is, for all $a, b, c \in \mathbb{Z}$, $a + (b + c) = (a + b) + c$.
- The number 0 is the *additive identity*; that is, for all $a \in \mathbb{Z}$, $0 + a = a + 0 = a$.
- Addition is *commutative*; that is, for all $a, b \in \mathbb{Z}$, $a + b = b + a$.
- For every element $a \in \mathbb{Z}$, there exists an element $-a$, *the inverse of a*, such that $a + (-a) = -a + a = 0$.

With respect to multiplication, we have:

- Multiplication is *associative*; that is, $a(bc) = (ab)c$.
- The number 1 is the *multiplicative identity*; that is, for all $a \in \mathbb{Z}$, $1 \cdot a = a \cdot 1 = a$.

We also have that for any $a \in \mathbb{Z}$, $0 \cdot a = a \cdot 0 = 0$ and $-a = (-1)a$. Multiplication is *commutative*; that is, $ab = ba$. Moreover, the *distributive law* holds; that is, for all $a, b, c \in \mathbb{Z}$, $a(b + c) = ab + ac$.

2.9. Using only the previous properties, prove the following statement that we first heard way back in elementary school: *A negative integer multiplied by a negative integer yields a positive integer.* (*Remark:* To be complete, we note that the numbers 1, 2, 3, ... are the *positive integers*, whereas −1, −2, −3, ... are the *negative integers.*)

The domino effect of bringing everyone down through guilt by induction

There is an important technique that enables us to prove infinitely many theorems in one proof. To inspire the basic idea, suppose a friend calls and tells us that she has set up a large number of dominos on their edges in an extremely elaborate and artful pattern.

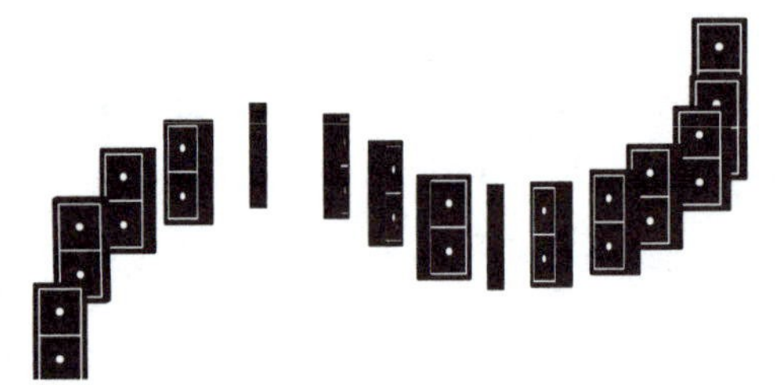

She also reports that if any particular domino falls, it will cause the next domino in line to fall.

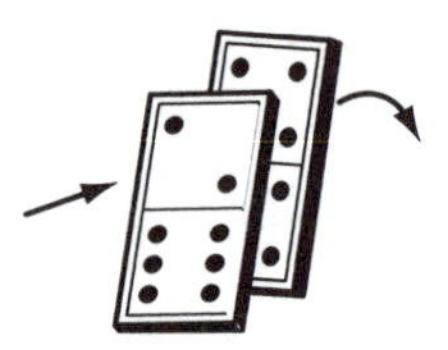

If one domino falls, then the next will also fall.

If she pushes over the first domino to cause it to fall, what can we conclude? We conclude that *all* the dominos will fall. Why? Because if we name a particular domino, then we see that given the facts at hand, it must fall at some point. For example, how can we prove that the fourth domino fell? Well, we were told that the first domino fell. Thus we know that it pushed over the second domino; which, in turn, pushed over the third domino; which, finally, toppled the fourth domino.

Given that the first domino falls—together with the general fact that if any particular domino falls, then the next domino also falls—we are able to prove that all the dominos go down. If we now view the collection of dominos as an infinite list of mathematical statements and replace the notion of "fallen" with "proven true," then we will arrive at the proof technique known as *induction*.

Specifically and more precisely, suppose we wish to establish the validity of a statement that is to hold for all natural numbers $n = 1, 2, 3, \ldots$. A proof by induction requires us to start by proving the truth of the statement in the first case (usually the case $n = 1$). This specific case is often referred to as the *base case* and is usually very easy to verify directly. Next, we assume the truth of the statement for an arbitrary integer, say $n = k$ (this assumption is known as the *inductive hypothesis*). We then use this assumption to deduce the truth of the statement in the next case—namely, for $n = k + 1$. That is, we use the assumed truth of the statement when $n = k$ to show that the analogous statement holds true for the next natural number, $k + 1$. Once we have successfully completed these two steps, the base case implies that the statement holds in the case $n = 1$. And because k was an arbitrary, unspecified natural number, we see that if the statement holds for any natural number, then it must also hold for the next natural number. Therefore we conclude that the original statement holds for all n by the domino effect described above—or, in this context, by mathematical induction.

We illustrate this important technique with two examples. We first establish the following result:

THEOREM. *For all natural numbers n,*

$$1+2+\cdots+n=\frac{n(n+1)}{2}. \tag{2.1}$$

We note that this result actually contains infinitely many theorems—one for each n. In particular, this theorem is equivalent to the following infinitely long list of results:

THEOREM 1.

$$1=\frac{1(1+1)}{2}.$$

THEOREM 2.

$$1+2=\frac{2(2+1)}{2}.$$

THEOREM 3.

$$1+2+3=\frac{3(3+1)}{2}.$$

$\vdots$

THEOREM 100.

$$1+2+3+\cdots+99+100=\frac{100(100+1)}{2}.$$

$\vdots$

Proof of the general theorem. We proceed by induction on the natural number n. We first consider the base case, $n = 1$. In this case, the left side of the expression in (2.1) equals 1. We also note that for $n = 1$, the right side of (2.1) equals

$$\frac{n(n+1)}{2} = \frac{1(1+1)}{2} = \frac{2}{2} = 1.$$

Thus we see that the identity in (2.1) holds for $n = 1$. Hence we have just established the base case. (In other words, we have just shown that the first domino falls!)

Next, we assume identity (2.1) holds for $n = k$. That is, we assume the following identity holds:

$$1 + 2 + \cdots + k = \frac{k(k+1)}{2}. \tag{2.2}$$

Using this induction hypothesis, we wish to establish that identity (2.1) holds in the next case: when $n = k + 1$. Thus, in this case, we consider the sum of the first $k + 1$ natural numbers and notice that we can express this as

$$1 + 2 + 3 + \cdots + k + (k+1) = (1 + 2 + 3 + \cdots + k) + (k+1).$$

We can now apply the induction hypothesis in identity (2.2) to the quantity within the first set of parentheses above to deduce

$$\begin{aligned} 1 + 2 + 3 + \cdots + k + (k+1) &= (1 + 2 + 3 + \cdots + k) + (k+1) \\ &= \left(\frac{k(k+1)}{2}\right) + (k+1) \\ &= \left(\frac{k(k+1)}{2}\right) + \frac{2(k+1)}{2} \\ &= \frac{k(k+1) + 2(k+1)}{2} = \frac{(k+1)(k+2)}{2}, \end{aligned}$$

which is precisely identity (2.1) in the case when $n = k + 1$. Thus by assuming the truth of the identity in the case $n = k$ (we assume that the kth domino falls), we deduce that the identity also holds in the next case: $n = k + 1$ (we see that the next domino must also fall). Thus we have established by induction that the identity holds for *all* n, which completes our proof.

Next we offer a geometric illustration involving induction. Specifically, we establish:

THEOREM. *Let n be a positive integer. If a circle is cut into arcs by n distinct diameters, then the total number of arcs would equal $2n$.*

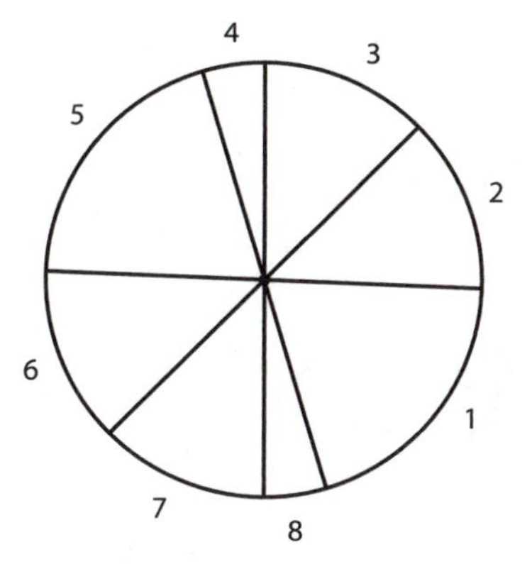

Four diameters cut the circle into eight arcs.

Proof. We proceed by induction on n. If $n = 1$, then we see that any one diameter partitions the circle into exactly two semicircles, and thus the number of arcs equals $2 = 2(1) = 2n$. Hence the statement holds for the base case $n = 1$.

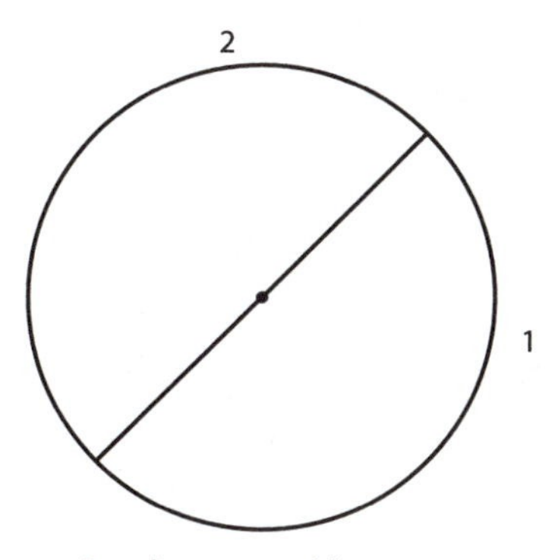

One diameter yields two arcs.

Next we assume that the theorem holds for $n = k$. That is, we assume that k distinct diameters partition the circle into exactly $2k$ arcs. We now consider $n = k + 1$ distinct diameters. If we momentarily ignore the $(k + 1)$th diameter, then we have k distinct diameters. By the inductive hypothesis, we know that these diameters partition the circle into $2k$ arcs.

Because the $(k + 1)$th diameter is distinct from the previous k diameters, we conclude that its endpoints must lie within the interior of two opposite arcs on the circle.

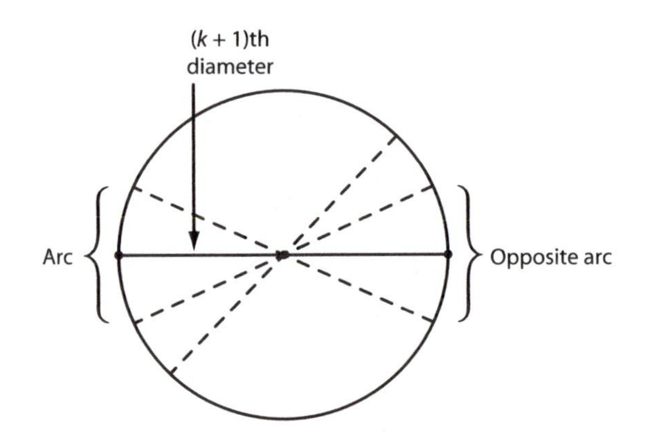

Endpoints of the $(k + 1)$th diameter lie inside a pair of opposite arcs.

Therefore we see that this last diameter partitions those opposite arcs each into two pieces. Those two arcs are now cut into four arcs, so we now have two additional arcs. Hence the total number of arcs in the circle partitioned by $k + 1$ diameters is $2k + 2$. Finally we observe that $2k + 2 = 2(k + 1) = 2n$, which establishes the result for the case $k + 1$. Therefore we see that the statement holds for all $n \geq 1$ by induction, which completes our proof.

Comment: Notice the writing style of the previous proofs. They are arguments consisting of paragraphs made up of sentences. They move forward in a logical and clear fashion. As you develop your mathematical voice, remember to express your ideas and arguments in clear, well-written prose. Again, be sure to visit Appendix 2 for a bit more commentary on crafting and writing proofs.

2.10. Prove *or* disprove the following:

STATEMENT. *For any integer $n \geq 1$,*

$$1+3+5+7+\cdots+(2n-1)=n^2.$$

2.11. Prove *or* disprove the following:

STATEMENT. *For any integer $n \geq 1$,*

$$1^2+2^2+3^2+\cdots+n^2=\frac{n(n+1)(2n+1)}{6}.$$

2.12. Prove *or* disprove the following:

STATEMENT. *For any integer $n \geq 2$,*

$$\left(1-\frac{1}{4}\right)\left(1-\frac{1}{9}\right)\left(1-\frac{1}{16}\right)\cdots\left(1-\frac{1}{n^2}\right)=\frac{n+1}{2n}.$$

Moreover, we have the formal identity

$$\prod_{n=2}^{\infty}\left(1-\frac{1}{n^2}\right)=\frac{1}{2}.$$

2.13. Prove *or* disprove the following:

STATEMENT. *Suppose the plane is divided into regions bounded by a finite number of distinct straight lines. Then each region can be painted either purple or gold in such a manner that any two regions sharing a common border edge will be painted different colors.*

(*Hint:* The spirit of the solution to this challenge mirrors one from Module 1.)

2.14. Prove *or* disprove the following:

STATEMENT. *Let the sequence of Fibonacci numbers, $\{F_n\}$, be defined by $F_0 = 1$, $F_1 = 1$, and for $n \geq 2$, $F_n = F_{n-1} + F_{n-2}$. Then for all $n \geq 0$,*

$$F_{n+2}F_n - (F_{n+1})^2 = (-1)^n .$$

Given positive numbers $a_0, a_1, a_2, \ldots$, we define two sequences, $\{p_n\}$ and $\{q_n\}$, as follows:

$$p_{-1} = 1,\ p_0 = a_0,\ \text{and for all } n \geq 1,\ p_n = a_n p_{n-1} + p_{n-2};\ \text{and}$$
$$q_{-1} = 0,\ q_0 = 1,\ \text{and for all } n \geq 1,\ q_n = a_n q_{n-1} + q_{n-2}.$$

2.15. Prove *or* disprove the following:

STATEMENT. *Let $\{p_n\}$ and $\{q_n\}$ be the sequences defined above. Then for all integers $n \geq 0$,*

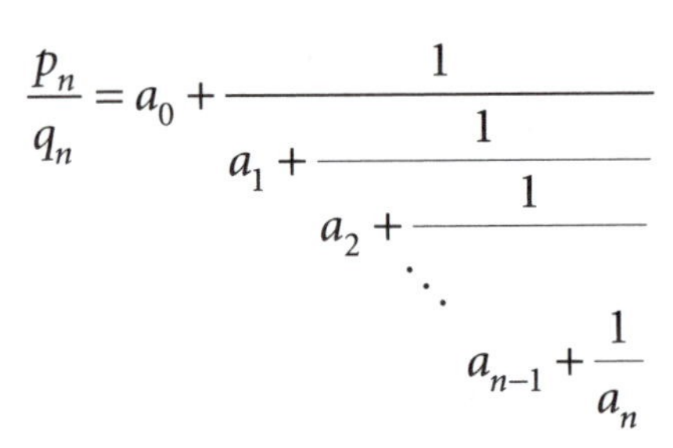

$$\frac{p_n}{q_n} = a_0 + \cfrac{1}{a_1 + \cfrac{1}{a_2 + \cfrac{1}{\ddots \, a_{n-1} + \cfrac{1}{a_n}}}}$$

Comment: The expansion in Statement 2.15 is called a *simple continued fraction* expansion. It is central in an area of number theory known as Diophantine analysis. These rational numbers lead to the "best" rational approximations to real numbers (such as $\frac{22}{7}$ for π).

Stepping back

Suppose you were asked to prove that there existed infinitely many numbers that exhibited a certain property. What proof techniques could you employ to establish such a result, and what would be the circumstances in which you might consider each technique?

Before moving ahead to the next module, you are encouraged to read or reread "A great adventure" from this book's introduction.

3

Delving into the dependable digits

Counting on counting numbers

Here we explore some of the wonderful ideas that form the foundation for the theory of numbers. We begin by recalling the set of *natural numbers*, written as $\mathbb{N}$ (a.k.a. the "counting" numbers)—$\mathbb{N} = \{1, 2, 3, \ldots\}$, and the set of *integers*, denoted by $\mathbb{Z}$, $\mathbb{Z} = \{\ldots, -3, -2, -1, 0, 1, 2, 3, \ldots\}$. Some may guess that we use $\mathbb{Z}$ for integers because it is just $\mathbb{N}$ laid on its right side, but others who know the German word for "number" might make a different conjecture.

Divide and conquer

Given two integers m and n, we say that m *divides* n, denoted $m \mid n$, if there exists an integer l such that $ml = n$. That is, m divides n if m is a factor of n. So, for example, $4 \mid 12$ because $4 \cdot 3 = 12$. (*Caution:* Note that $m \mid n$ is very different from m/n. For example, $m \mid 0$, but we would never dare write that $m/0$.) If m does not divide n, then we write $m \nmid n$. An integer c is said to be a *common divisor* of the integers a and b if $c \mid a$ and $c \mid b$ (sometimes c is called a *common factor* of a and b). We say that a positive integer g is the *greatest common divisor* of a and b if g is a common divisor of both a and b, with the additional property that if c is any other common divisor of a and b, then $c \mid g$. We denote the greatest common divisor of a and b as $\gcd(a, b)$.

3.1. Compute, by hand, the gcd(168, 180). Would you enjoy the arithmetical challenge of finding the gcd(3913, 23177) by hand? (Just give a short answer: Yes *or* No—no justification is required here.) Then prove the following very useful fact: For integers a, b, d, m, and n, if $d \mid m$ and $d \mid n$, then $d \mid (am + bn)$. That is, if d divides both m and n, then d divides any linear combination of m and n having integer coefficients.

3.2. Prove the following:

THEOREM. *Suppose that m, n, q, and r are integers satisfying the identity $n = mq + r$. Then* $\gcd(m, n) = \gcd(m, r)$.

Thus if r is *smaller* than n, then Theorem 3.2 allows us to replace the potentially daunting task of computing the $\gcd(m, n)$ with the slightly simpler task of finding the $\gcd(m, r)$. Of course, if we could repeat this reduction process, then we could turn a challenging problem into child's play. We now explore this possibility.

3.3. Prove and extend *or* disprove and salvage:

STATEMENT. *Given natural numbers m and n, there exists a unique pair of integers q and r such that $n = mq + r$ and $0 \leq r < m$.*

3.4. Prove the following:

THEOREM (The Division Algorithm). *Let m and n be two integers with $m \neq 0$. Then there exists a unique pair of integers q and r satisfying*

$$n = mq + r \quad \textit{and} \quad 0 \leq r < |m|.$$

In the Division Algorithm, q is known as the *quotient* and r is referred to as the *remainder.* In fact, the Division Algorithm is the formalization of what was called "long division" long ago in the carefree (and proof-free) days of our youth.

Solving equations by simple division

One of the most useful applications of the Division Algorithm is in finding integer solutions to certain linear equations.

3.5. Prove the following:

THEOREM (The Euclidean Algorithm). *Let m and n be integers with $m \neq 0$. Then there exists a finite collection of pairs of integers $(q_1, r_1), (q_2, r_2), \ldots, (q_L, r_L)$ satisfying*

$$\begin{aligned}
n &= mq_1 + r_1 && \text{and} \quad 0 < r_1 < |m|, \\
m &= r_1 q_2 + r_2 && \text{and} \quad 0 < r_2 < r_1, \\
r_1 &= r_2 q_3 + r_3 && \text{and} \quad 0 < r_3 < r_2, \\
r_2 &= r_3 q_4 + r_4 && \text{and} \quad 0 < r_4 < r_3, \\
&\ \vdots && \qquad\quad \vdots \\
r_{L-3} &= r_{L-2} q_{L-1} + r_{L-1} && \text{and} \quad 0 < r_{L-1} < r_{L-2}, \\
r_{L-2} &= r_{L-1} q_L + r_L && \text{and} \quad r_L = 0.
\end{aligned}$$

Moreover, $\gcd(m, n) = r_{L-1}$.

3.6. Using the Euclidean algorithm, compute the gcd(3913, 23177) by hand. Was the process as arithmetically challenging as you initially guessed in Challenge 3.1?

3.7. Using the identities found in Challenge 3.6, find integers x and y that satisfy the following linear equation:

$$3913x + 23177y = 301.$$

(No guess-and-checking allowed! Find a systematic method using your work from Challenge 3.6.)

The linear equation in Challenge 3.7 is an example of what is known as a *linear Diophantine equation*. The word "Diophantine" honors the Greek mathematician Diophantus, who was interested in finding *integer* solutions to various sundry equations way back in the olden days (around the third century A.D.).

3.8. Prove and extend *or* disprove and salvage (with another "if and only if" statement):

STATEMENT. *Given integers m, n, and g, the integer g equals the* gcd(m, n) *if and only if there exist integers x and y satisfying the linear Diophantine equation*

$$mx + ny = g.$$

If gcd(a, b) = 1, then we say that a and b are *relatively prime*; that is, they share no positive common factors other than 1.

3.9. Prove and extend *or* disprove and salvage (with another "if and only if" statement):

STATEMENT. *Two integers m and n are relatively prime if and only if there exist integers x and y satisfying the linear Diophantine equation*

$$mx + ny = 1.$$

The theorem found in the previous challenge is extremely useful. In fact, whenever we are told that two integers are relatively prime, our first course of action should be to apply the theorem we have just discovered. Let's illustrate the theorem's utility with a couple of useful applications. (Check out the cover of this book for an attempt at 3.10(a).)

3.10. (a) Prove and extend *or* disprove and salvage:

STATEMENT. *Let k, m, and n be integers. If $k \mid mn$ but $k \nmid m$, then $k \mid n$.*

3.10. (b) Prove and extend *or* disprove and salvage:

STATEMENT. *Let a, b, and m be integers. If a and b are relatively prime and $a \mid m$ and $b \mid m$, then $ab \mid m$.*

Now ready for prime time

An integer $p > 1$ is said to be *prime* if the only positive integers m that divide p are $m = 1$ and $m = p$. An integer $c > 1$ is called *composite* if it is not prime. So we note the quaint convention that 1 is neither prime nor composite. (Actually this is neither *quaint* nor a *convention*. There is a reason for this custom, which becomes clearer with more advanced mathematics. Basically, 1 is neither prime nor composite because it possesses a special property—its multiplicative inverse (namely, itself) is also an integer. Of course, we know that 1 is the only natural number with this property. If the multiplicative inverse of 2 (which we know to be $\frac{1}{2}$) *were* an integer, then 2 would be neither prime nor composite as well. Fun, huh?)

3.11. Prove and extend *or* disprove and salvage:

STATEMENT. *For a prime number $p > 5$, the number $p^2 - 1$ is a multiple of* 12.

3.12. Prove and extend *or* disprove and salvage:

STATEMENT. *Let m and n be integers. Suppose p is a prime number such that $p \mid mn$. Then either $p \mid m$ or $p \mid n$.*

3.13. Prove the following:

THEOREM (The Fundamental Theorem of Arithmetic). *Every integer $n > 1$ can be expressed as a finite product of prime numbers. Moreover, that product is unique, except for possible reorderings of the factors.*

3.14. Prove and extend *or* disprove and salvage:

STATEMENT. *If m and n are integers with $m > 1$ and $m \mid n$, then $m \nmid n + 1$.*

3.15. Prove the following theorem due to Euclid. This result is considered by many to be one of the most important and beautiful theorems in mathematics.

THEOREM. *There are infinitely many prime numbers.*

Stepping back

As we have discovered, the unique factorization property of the natural numbers allows us to prove several foundational results. Which theorems required unique factorization? Can you imagine a collection of numbers that does not exhibit unique factorization? That is, can you find a collection of numbers C such that every number in C can be factored into "prime" elements *from the collection* C but in such a way that the factorization might not be unique?

4

Going around in circles

The art of modular arithmetic

Here we discover that many arithmetical notions can be converted into the powerful language of congruences—a world where integers are equivalent if they have similar divisibility properties. While such a world might at first appear foreign, keep in mind that we use such cyclical arithmetic whenever we tell time or determine the day of the week.

The mod side of mathematics

Let a, b, and m be three integers with $m > 1$. We say that *a is congruent to b modulo m*, denoted $a \equiv b \bmod m$, if $m|(a - b)$. The integer m is called the *modulus*, and often we read the relation $a \equiv b \bmod m$ as *a is equivalent to b mod m*. Note that in some sense, on a 12-hour clock, 16 o'clock is the same time numerically as 4 o'clock (that is, 16 o'clock $\equiv$ 4 o'clock mod 12) because $12|(16 - 4)$. Similarly, in terms of days of the week, June 2 $\equiv$ June 16 mod 7 because $7|(2 - 16)$.

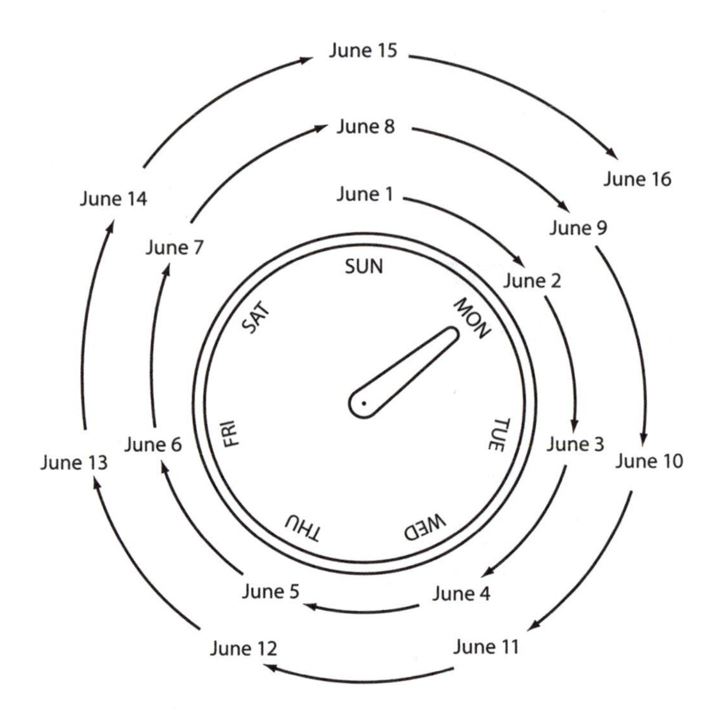

"Day of the Week" Clock (mod 7)

4.1. Using the definition of "congruence," verify that the notion of congruence gives rise to an *equivalence relation*; that is, verify the following properties for all integers a, b, c, and all $m > 1$:

Reflexive: $a \equiv a \bmod m$.

Symmetric: If $a \equiv b \bmod m$, then $b \equiv a \bmod m$.

Transitive: If $a \equiv b \bmod m$ and $b \equiv c \bmod m$, then $a \equiv c \bmod m$.

4.2. Prove and extend *or* disprove and salvage (with another "if and only if" statement):

Statement. *Let a, b, and m be integers with $m > 1$. Then* $a \equiv b \bmod m$ *if and only if the remainder when a is divided by m equals the remainder when b is divided by m.*

4.3. Prove and extend *or* disprove and salvage:

Statement. *Let $m > 1$ be an integer and let $S = \{0, 1, 2, \ldots, m - 1\}$. Then given any integer a, there exists a unique integer $r \in S$ such that* $a \equiv r \bmod m$.

Any set S that satisfies the conclusion of Statement 4.3 is called a *complete residue system modulo m*. Thus, in view of the previous result, we see that the set $\{0, 1, 2, \dots, m-1\}$ is a complete residue system.

Old-school algebra gets an extreme makeover

We now explore some of the arithmetical and algebraic properties that congruence relationships possess. Most of the properties that we know and love involving equalities carry over to the world of congruences. But do *all* such properties hold in this new setting? Let's see ...

4.4. Prove and extend *or* disprove and salvage:

STATEMENT. *Let a, b, c, c′, and m be integers with* $m > 1$. *Suppose that* $c \equiv c' \bmod m$. *If* $a \equiv b \bmod m$, *then*

$$(a + c) \equiv (b + c') \bmod m \quad \text{and} \quad ac \equiv bc' \bmod m.$$

4.5. Prove and extend *or* disprove and salvage (with another "if and only if" statement):

STATEMENT. *Let a, b, c, and m be integers with* $m > 1$. *Then*

$$ac \equiv bc \bmod m \quad \text{if and only if} \quad a \equiv b \bmod m.$$

4.6. Given integers a, b, and m, with $m > 1$, find a necessary and sufficient condition involving only a, b, and m to ensure that there exists an integer solution x to the linear congruence $ax \equiv b \bmod m$.

4.7. Prove and extend *or* disprove and salvage (with another "if and only if" statement):

STATEMENT. *Let* m_1 *and* m_2 *be two relatively prime integers each greater than* 1. *Then for integers a and b, the system of simultaneous linear congruences*

$$\begin{aligned} ax &\equiv b \bmod m_1 \\ ax &\equiv b \bmod m_2 \end{aligned}$$

has a solution $x \in \mathbb{Z}$ *if and only if there is an integer solution to the linear congruence*

$$ax \equiv b \bmod m_1 m_2.$$

The power of congruences

Here we discover a powerful result of Fermat that provides a valuable insight into the structure of congruences modulo a prime number. As we will see for ourselves later in this module, Fermat's ancient result, which is more than 350 years old, is the key to allowing us to securely transfer information in our modern world.

4.8. Prove and extend *or* disprove and salvage:

STATEMENT. *Suppose that a and m are relatively prime integers with $m > 1$. Then the integers $a, 2a, 3a, \dots, (m-1)a$ are distinct modulo m.*

4.9. If p is a prime number and a is an integer relatively prime to p, then compare the two sets

$$\{1, 2, \dots, p-1\} \text{ modulo } p \quad \text{and} \quad \{a, 2a, \dots, (p-1)a\} \text{ modulo } p.$$

4.10. Prove the following:

THEOREM (Fermat's Little Theorem). *Let p be a prime and a be an integer relatively prime to p. Then*

$$a^{p-1} \equiv 1 \bmod p.$$

4.11. Prove and extend *or* disprove and salvage:

STATEMENT. *Let a and m be two integers with $m > 1$. Then*

$$a^m \equiv a \bmod m.$$

4.12. Establish the following result (which makes a terrific bar bet):

Let n be *any* integer. Then 15 divides $11n^8 + 4n^4$.

(*Bar-bet version:* "I'll bet you \$5.00 that for any big counting number n you think of, I can find a factor of $11n^8 + 4n^4$ other than itself and 1 without your telling me the number n." *Caution:* You might get beat up if you actually try this money-making venture.)

Public secret codes

If you wish to send a confidential message, you must first *encode* the message and then send the encrypted message so the receiver can *decode* it to learn of your deepest secrets. It may seem intuitively obvious that the encoding scheme itself must be one of your tightly kept secrets; otherwise your coded message could be intercepted and read by less desirable people (you know who they are). Here we will discover that, paradoxically, the most convenient way to keep your coded messages safe from eavesdroppers is to make everyone's coding schemes completely *public* for all to know. This counterintuitive encryption scheme is known as a *public key code* and is now used millions of times daily for Internet communication and ATM banking. We now consider the wonderful 1977 discovery of Rivest, Shamir, and Adleman that gave rise to the popular public key coding scheme known as RSA.

We first face the obvious question: If everyone knows how to encode a message to an individual, then is the message really a secret anymore? The basic theme of the "public" aspect of the RSA coding scheme is given in the following ten sentences from *The Heart of Mathematics: An Invitation to Effective Thinking,* by E.B. Burger and M. Starbird:

> Let's select two enormous prime numbers—and we mean enormous—say each having about 300 digits—and multiply those numbers together. How can we multiply them together? Computers are whizzes at *multiplying* natural numbers—even obscenely long ones. *Factoring* larger numbers, however, is hard—even for computers. Computers are smart but not infinitely smart—there are limits to the size of natural numbers that they can factor. In fact, our product is much too large for even the best computers to factor. So if we announce that huge product to the world, even though it can be factored in theory; in practice it cannot. Thus we are able to announce the gigantic number to everyone, and yet no one but we would know its two factors. This huge product is the *public* part of the RSA public key code. Somehow, the fact that only we know how to factor that number allows us to decode messages while others cannot.

To set up your own RSA public key code, first select two *distinct* prime numbers, p and q, and define $n = pq$ and $m = (p - 1)(q - 1)$. Next choose a positive integer e that is relatively prime to m and assume you can find integers

y and d, with $d > 0$, satisfying the relation $ed = 1 + my$. Keep the integer d (the *decoding* integer) secret; announce the pair of integers (n, e) to the public (e is the *encoding* integer); and destroy all documents and files containing the integers m, p, q, and y. Your public code is now ready for use.

To illustrate this one-time setup, let's create your RSA encryption scheme, using ridiculously tiny numbers. Let's pick $p = 3$ and $q = 5$, so $n = 15$. We now let $m = (3 - 1)(5 - 1) = 8$ and select e to be any number relatively prime to 8; let's choose $e = 11$. We now wish to find integer solutions d and y, with $d > 0$, to the linear equation $11d = 1 + 8y$. We find that one such solution is $d = 3$ and $y = 4$. Thus you announce the pair of numbers $(n, e) = (15, 11)$ to everyone—your friends, family, and even your unsavory enemies. Those are the two numbers needed if anyone wishes to send you a coded message. You keep $d = 3$ a secret. You tell it to *no one*. Not even your closest friend. If in a romantic moment, a special someone whispers into your ear, "Tell me your d," you should jump up and say, "Never!" The other numbers (m, p, q, and y) are no longer needed and, for security reasons, should be destroyed. Of course, this choice of n is silly, because almost everyone can factor it, but we're just trying to illustrate the method.

Suppose Alice wishes to send you a message. She would first convert each letter into a number by simply substituting A = 01, B = 02, . . . , Z = 26, so that her message would be a very long natural number (for example, HI would become 0809). Then she would cut that number into small "words"—integers less than n (recall that both e and n were made public). So in our example above, because you have $n = 15$, and 0809 is larger than 15, Alice would break it up into two "words": 08 and 09. Suppose now that W is one of these words—that is, W is a positive integer such that $W < n$. Alice then computes the remainder C when W^e is divided by n. The number C is the *coded* version of the word W. You receive the word C and are ready to decode it. Given the word C, with $C < n$, you find the remainder D when C^d is divided by n (recall that only you know d). You follow this process for each individual "word," then string them together and convert the numbers back to letters to read Alice's original message.

An explicit numerical example

Returning to our previous example, for Alice to encode 08, she first reduces 8^{11} mod 15. Here we offer one of many possible ways to compute the remainder by hand:

$$8^{11} = (8^2)^5(8) \equiv (4)^5(8) = (16)^2(4)(8) \equiv (1)^2(4)(8) \equiv 2 \bmod 15.$$

Thus 2 is the coded version of 8. Alice announces 2 on the Web, in blogs, and in newspaper ads for all to see. However, you are the only one with the secret $d = 3$ to decode it. When you receive the coded message 2, you find the remainder when $2^d = 2^3$ is divided by 15. In this case, it's trivial: $2^3 = 8$, so the remainder is 8. Notice that 8 is Alice's original message! Is this cool coincidence magic or mathematics?

4.13. Show that given the notation above, there will always exist integers y and d, with $d > 0$, satisfying $ed = 1 + my$.

4.14. Given the notation above, prove that $D = W$; that is, show that the decoded message D is the same as the original message W.

4.15. Of course, given that everyone's encoding scheme is public, it would be trivial for Bob to send a message to us and assert that it is from Alice. Can you devise a means of creating an electronic "signature" where we can know with certainty that a message we receive from Alice is actually from her?

Stepping back

Suppose you were able to factor the integer n. Describe a method for breaking that RSA code and show that your procedure would indeed crack the code.

5

The irrational side of numbers

A world of nonrepeating digits

In our daily lives, we see that everything around us can be measured with fractions. Even the price of gasoline conveniently ends with the fraction $\frac{9}{10}$ of a cent (lucky us!). Given our life experience, it is completely natural, if not utterly obvious, to believe that all numbers are fractions. In fact, this intuitive-sounding assertion goes back to the ancient Greeks. Their entire worldview embraced this rational notion of number. Here we will see that this dreamy vision of a numerically rational worldview is just that: a dream.

Divisible and indivisible

When we consider quotients of integers, we immediately come face-to-face with fractions. We define the set of *rational numbers,* denoted $\mathbb{Q}$, to be

$$\mathbb{Q} = \left\{ \frac{r}{s} : r, s \in \mathbb{Z}, s \neq 0 \right\}.$$

Why $\mathbb{Q}$? Perhaps in honor of *quotients.*

5.1. If a and b are nonzero integers, can the number

$$\frac{a}{b} + \frac{b}{a}$$

ever be an integer? If not, prove why not; if so, find all such integers a and b.

The set of *real numbers,* denoted $\mathbb{R}$, can be viewed informally as the collection of all points on a number line. For thousands of years, it was believed that every point on the number line was rational. People intuitively "knew" that the set of real numbers and the set of rational numbers were the same set. The discovery that these sets are not equal was a tremendous and dramatic breakthrough. An *irrational number* is a number that is not rational. The discovery of irrational numbers was one of the great triumphs of mathematics.

5.2. Prove and extend *or* disprove and salvage:

STATEMENT. *There do not exist integers m and n satisfying* $m^2 = 2n^2$.

5.3. Prove that $\sqrt{2}$ is irrational.

5.4. Prove that $\sqrt{p}$ is irrational for any prime number p. Apply your argument to the number $\sqrt{4}$ and show where your proof breaks down (and we know it must, because $\sqrt{4}$ is not very irrational).

Forgoing fractions

Now that we have discovered for ourselves the existence of numbers that are not rational, we are ready to consider some attractive and challenging statements.

5.5. Prove that if $m > 1$ is an integer with the property $\sqrt{m} \in \mathbb{Q}$, then $\sqrt{m} \in \mathbb{Z}$. Then use the contrapositive of the result to *easily* establish the irrationality of $\sqrt{35}$.

5.6. Prove and extend *or* disprove and salvage:

STATEMENT. *Let* $F_0 = 1$, $F_1 = 1$, $F_2 = 2$, $F_3 = 3$, $F_4 = 5$, … *denote the Fibonacci numbers. Then for* $n > 1$, $\sqrt{F_n F_{n+2}}$ *is irrational.*

(*Follow-up:* Does a similar result hold for the number $\sqrt{F_n}$?)

5.7. Prove *or* disprove:

STATEMENT. *If τ_1 and τ_2 are irrational numbers, then $\tau_1^{\tau_2}$ is irrational.*

(*Hint:* Take $\tau_2 = \sqrt{2}$ and consider two possibilities for τ_1.)

Irrationalit-*e*

Recall from your old calculus daze that $e = \sum_{n=0}^{\infty} \frac{1}{n!}$. This number is one of the most important constants in mathematics.

5.8. *Without* using the decimal expansion for *e*, prove that

$$0 < \sum_{n=2}^{\infty} \frac{1}{n!} < 1.$$

5.9. Prove that *e* is not an integer.

5.10. Prove that *e* is irrational.

Stepping back

Find a necessary and sufficient condition on the digits of the decimal expansion of a real number to ensure that it is rational. Using your answer, determine whether $n.9999\ldots$ is rational, where $n \in \mathbb{N}$, and whether $\sum_{n=1}^{\infty} 10^{-n^2}$ is rational. In both cases, if the number is rational, express it as a fraction. Using your answer as inspiration, consider the following question: If a person were to select a real number at random (perhaps placing a pin on the real number line while blindfolded), what is the likelihood that the number selected is irrational?

6

Discovering how to function in mathematics

Moving beyond ordinary relations

Here we explore some of the basic notions surrounding two of the most fundamental objects of mathematics: sets and functions. These two players are central to all of mathematics and star in nearly every intellectual pursuit beyond mathematics. For the computer generation, a function can be viewed as a machine or program that when given an input, produces an output. Functions allow us to establish important connections between collections and to explore causality. With these connections, we can uncover hidden structure and patterns within mathematics, nature, and our lives.

Get ready, get set

By a *set* we intuitively mean a collection of elements. The *empty set,* denoted $\emptyset$, is the set that contains no elements. If a is an element of the set A, we write $a \in A$. Given sets A and B, we define the *union of A and B* to be the set given by

$$A \bigcup B = \{s : \text{either } s \in A \text{ or } s \in B\}$$

and the *intersection of A and B* to be the set given by

$$A \bigcap B = \{s : s \in A \text{ and } s \in B\}.$$

(We often express sets using set-builder notation. So the set $\{s : s$ satisfies a certain condition$\}$ is read, "The set of all s such that s satisfies a certain condition." Furthermore, if $a \in \{x : x$ satisfies condition $C\}$, then we know that a satisfies condition C.) Two sets are *equal* if they contain the same elements. For example, $\{1, 2, 3\} = \{3, 1, 1, 2, 1\}$; that is, order and multiplicity are irrelevant.

We say that A and B are *disjoint sets* if they share no elements in common; that is, $A \cap B = \emptyset$. We say that A is a *subset* of B, denoted $A \subseteq B$, if every element of A is an element of B. We say that A is a *proper subset* of B if $A \subseteq B$ but $A \neq B$. To establish that $A \subseteq B$, we usually consider an arbitrary element $a \in A$ and demonstrate that, in fact, $a \in B$; that is, every element of A is an element of B. Thus we note that $A = B$ if and only if $A \subseteq B$ *and* $B \subseteq A$. Often in mathematics, when we wish to establish that two sets are equal, we prove that the previous two set containments hold. Finally, we define the set *A minus B*, written $A \setminus B$, by

$$A \setminus B = \{a \in A : a \notin B\}$$

(read: "The set of all a in A such that a is not in B").

6.1. For *any* sets A and B, which of the following statements are true? Verify the ones that are true and salvage the ones that are false.

- $\emptyset \in A$
- $A \subseteq A$
- $A \cap B \subseteq A \cup B$
- $(A \setminus B) \setminus C = A \setminus (B \setminus C)$
- $A \cup (B \cap C) = (A \cap B) \cup (A \cap C)$
- $A \setminus B \subseteq A$
- $A \subseteq B$ if and only if $A \cup B = B$

Given two sets S_1 and S_2, we define the *Cartesian product of S_1 and S_2*, denoted $S_1 \times S_2$, to be the set

$$S_1 \times S_2 = \{(s_1, s_2) : s_1 \in S_1 \text{ and } s_2 \in S_2\}.$$

Thus $S_1 \times S_2$ is the set of all *ordered* pairs in which the first element is from S_1 and the second is from S_2. For example, if $S_1 = \{a, b\}$ and $S_2 = \{1, 2, 3, 4\}$, then $(b, 4)$ is an element of the set $S_1 \times S_2$. In fact,

$$S_1 \times S_2 = \{(a, 1), (a, 2), (a, 3), (a, 4), (b, 1), (b, 2), (b, 3), (b, 4)\},$$

and we note that $(3, a) \notin S_1 \times S_2$, whereas $(b, 4) \in S_1 \times S_2$, as we said b4 (sorry). Given a set S and a positive integer n, we write S^n for the Cartesian product given by

$$S^n = \underbrace{S \times S \times S \times \cdots \times S}_{n \text{ copies of } S}.$$

Given a set S, we define the *cardinality of* S, denoted $|S|$, to be the number of elements contained in S. Note that duplicate elements are not counted (because multiplicity is irrelevant for sets). In other words, $S = \{0, 0, 0\}$ contains just one element; hence, $|S| = 1$. It follows that if S_1 and S_2 are two nonempty, finite sets, then

$$|S_1 \times S_2| = |S_1||S_2|.$$

Fun with functions

Let X and Y be two nonempty sets. A *function* $f : X \to Y$ is a subset of the Cartesian product $X \times Y$ with the property that for each $x \in X$, there exists exactly one $y \in Y$ satisfying $(x, y) \in f$. The set X is called the *domain of* f, and the set Y is called the *codomain* of f.

Returning to the sets $S_1 = \{a, b\}$ and $S_2 = \{1, 2, 3, 4\}$ from b4 (that will be the last time—promise), we see that

$$f = \{(a, 1), (b, 1)\} \quad \text{and} \quad g = \{(a, 3), (b, 2)\}$$

are examples of functions, while $h = \{(a, 2), (a, 3)\}$ is not.

6.2. Prove and extend *or* disprove and salvage:

STATEMENT. *Let $f : X \to Y$ be a function, and suppose that $x_0 \in X$ and $y_0 \in Y$ satisfy $(x_0, y_0) \in f$. If $(x, y) \in f$ and $x \neq x_0$, then $y \neq y_0$.*

If $S \subseteq X$, then we define the set $f(S)$ to be the subset of Y given by

$$f(S) = \{y \in Y : \text{there exists an } s \in S \text{ such that } (s, y) \in f\}.$$

The set $f(S)$ is called the *image of the set S under the function f.* We define the *range of f* to be the set $f(X)$. Sometimes the range of f is called the *image of f.* Returning to our previous examples, we see that $f(S_1) = \{1\}$ and $g(S_1) = \{2, 3\}$.

We simplify our notation by viewing a function f as a *map* from X into Y. Moreover we now write $y = f(x)$ to denote that $(x, y) \in f$. Thus for any subset S of X, we note that

$$f(S) = \{f(s) : s \in S\}.$$

So in the case of our previous examples, we could write $f(a) = 1, f(b) = 1$ and $g(a) = 3, g(b) = 2$.

6.3. If X and Y are finite sets and $f: X \to Y$ is a function, then what is the relationship between $|f(X)|$ and $|Y|$? How does $|f(X)|$ compare with $|X|$?

If $f: X \to Y$ and $g: Y \to Z$ are functions, we define *f composed with g*, written $g \circ f$, to be the subset of $X \times Z$ given by

$$g \circ f = \{(x, z) \in X \times Z : \text{there exists an element } y \in Y \text{ satisfying } (x, y) \in f \text{ and } (y, z) \in g\}.$$

6.4. Prove and extend *or* disprove and salvage:

STATEMENT. *If $f: X \to Y$ and $g: Y \to Z$ are functions, then the composition $g \circ f$ is also a function. In particular, $g \circ f: X \to Z$, and $(g \circ f)(x) = g(f(x))$.*

Moving onto an intimate one-to-one relationship with functions

We say that a function $f: X \to Y$ is *one-to-one* if whenever $(x_1, y) \in f$ and $(x_2, y) \in f$, it then follows that $x_1 = x_2$. That is, if $f(x_1) = f(x_2)$, then x_1 must equal x_2. We say that a function $f: X \to Y$ is *onto* if for every $y_0 \in Y$, there exists an $x_0 \in X$ such that $(x_0, y_0) \in f$; that is, given $y_0 \in Y$, there exists an $x_0 \in X$ such that $f(x_0) = y_0$. Thus if f is onto, then $f(X) = Y$. (Can you prove it? Try!)

6.5. Produce explicit functions $f: \mathbb{R} \to \mathbb{R}$ such that

- f is one-to-one and onto.
- f is one-to-one but not onto.
- f is onto but not one-to-one.
- f is neither one-to-one nor onto.

6.6. Suppose that $f: X \to Y$ and $g: Y \to Z$ are functions. If f and g are each one-to-one, then must $g \circ f$ also be one-to-one? If f and g are each onto, then must $g \circ f$ also be onto?

Suppose that $f: X \to Y$ is a function. We now define the *inverse of* f, denoted f^{-1}, to be the subset of $f(X) \times X$ given by

$$f^{-1} = \{(y, x) \in Y \times X : (x, y) \in f\}.$$

6.7. Prove and extend *or* disprove and salvage:

STATEMENT. *Given a function $f: X \to Y$, its inverse $f^{-1}: f(X) \to X$ is also a function. Moreover for all $y \in f(X)$, $(f \circ f^{-1})(y) = y$, and for all $x \in X$, $(f^{-1} \circ f)(x) = x$.*

For any function $f: X \to Y$ and subset $R \subseteq Y$, we define the set

$$f^{-1}(R) = \{x \in X : \text{there exists a } y \in R \text{ such that } f(x) = y\}.$$

The set $f^{-1}(R)$ is called the *preimage of R*.

6.8. Prove and extend *or* disprove and salvage (with another "if and only if" statement):

STATEMENT. *A one-to-one function $f: X \to Y$ is onto if and only if $|X| = |Y|$.*

6.9. Let $2\mathbb{Z}$ denote the set of all even integers. Can there exist a one-to-one, onto function $f: \mathbb{Z} \to 2\mathbb{Z}$?

6.10. Prove and extend *or* disprove and salvage (with another "if and only if" statement):

STATEMENT. *Two finite sets A and B have the same cardinality if and only if there exists a one-to-one function $f: A \to B$.*

Given a set A, we define the *power set of A*, denoted $\mathcal{P}(A)$, to be the set of all subsets of A.

6.11. List all the elements in the set $\mathcal{P}(\{\clubsuit, \diamondsuit, \heartsuit, \spadesuit\})$.

6.12. Suppose that $|A| = n$. Construct a one-to-one function $f: \mathcal{P}(A) \to \{0, 1\}^n$. Is your map onto?

6.13. Prove and extend *or* disprove and salvage:

STATEMENT. *Given a finite set A, $|\mathcal{P}(A)| = 2^{|A|}$.*

An intuitively-believable-yet-challenging-to-verify correspondence

Suppose that for two sets A and B, there exist two one-to-one functions, $f: A \to B$ and $g: B \to A$. Intuitively it may seem apparent that there must exist a function $h: A \to B$ that is both one-to-one *and* onto. This important result is known as the Schroeder-Bernstein Theorem.

6.14. Prove the following:

THEOREM (Schroeder-Bernstein Theorem). *Let A and B be two sets. If there exist one-to-one functions $f: A \to B$ and $g: B \to A$, then there exists a function $h: A \to B$ that is both one-to-one and onto.*

(*Hint:* First we make the following obvious observation: Given an element $a \in A$, there are only two natural functions we can use to define $h(a)$—either

f or g^{-1}. Now let us, just for visualization purposes, depict the sets A and B as vertical line segments. Thus we have a picture that resembles

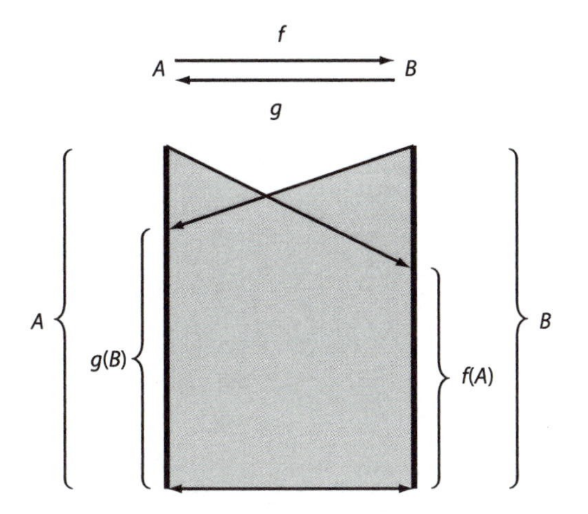

For which elements $a \in A$ *must* we define h by $h(a) = f(a)$? What is the image of all such a's? What elements in A have yet to be mapped? If we want to ensure that h is one-to-one, what subset should we now consider? Continue this process from here to ...)

6.15. Describe a one-to-one, onto function from the closed interval [0, 1] onto the half-closed interval [0, 1).

Stepping back

Suppose that $f: \mathbb{Z} \to \mathbb{R}$ is a function that satisfies the following property: For any n and m in $\mathbb{Z}$, $f(n+m) = f(n) + f(m)$. Suppose we know the values $f(t)$ for all *positive, even* integers t. Would we know the value of the function f for *all* integers n? What is the smallest number of integers required so that if we knew the value of f at those integers, we would know the function for *all* integers?

7

Infinity

Understanding the unending

One of the greatest triumphs of human thought was the taming of infinity. In the late nineteenth century, Georg Cantor was the first to discover some of the many mysterious surprises surrounding infinity. Cantor conquered infinity by carefully crafting a simple yet precise definition of "equally numerous." The counterintuitive consequences that followed from his basic idea changed the course of mathematics forever.

Comparing cardinality

Let A and B be two sets. We say that A and B have *the same cardinality,* written $|A| = |B|$, if there exists a one-to-one, onto function $f: A \to B$. Now we can be more precise in our definition of a finite set: We say that a nonempty set S is *finite* if there exists a positive integer n such that $|S| = |\{1, 2, \ldots, n\}|$. A nonempty set S is *infinite* if it is not a finite set. Cantor defined a set S to be *infinite* if there exists a one-to-one, onto function from S to a *proper* subset of S. It can be shown that these two definitions are equivalent. We say that the empty set has cardinality 0 and thus declare it to be a finite set as well. (For the following challenges, recall that $\mathbb{N}$ denotes the set of natural numbers.)

7.1. Prove the following:

THEOREM. *The set $\mathbb{N}$ is an infinite set.*

7.2. Prove and extend *or* disprove and salvage:

STATEMENT. *The set $\{0\} \cup \mathbb{N}$ has the same cardinality as $\mathbb{N}$.*

7.3. Prove and extend *or* disprove and salvage:

STATEMENT. *The set $\mathbb{Z}$ has the same cardinality as $\mathbb{N}$.*

7.4. Prove and extend *or* disprove and salvage:

STATEMENT. *The set $\mathbb{Q}$ has the same cardinality as $\mathbb{N}$.*

Countable collections

A set S is said to be *countable* if either it is finite or it has the same cardinality as $\mathbb{N}$. We say that S is *countably infinite* if it is an infinite, countable set.

7.5. Prove and extend *or* disprove and salvage:

STATEMENT. *If S_1 and S_2 are two countable sets, then the set $S_1 \times S_2$ is countable.*

7.6. Given a natural number N, let $\mathcal{B}_N$ be the set defined by $\mathcal{B}_N = \{0,1\}^N$. That is,

$$\mathcal{B}_N = \{(x_1, x_2, \ldots, x_N) : x_n \in \{0,1\}\}.$$

Without *explicitly computing the cardinality of any sets,* prove that

$$|\{1, 2, \ldots, N\}| \neq |\mathcal{B}_N|.$$

Which of these sets is larger? What is the connection between the sets $\{1, 2, \ldots, N\}$ and $\mathcal{B}_N$?

7.7. We define the set $\mathcal{B}_\infty$ by

$$\mathcal{B}_\infty = \{(x_1, x_2, \ldots) : x_n \in \{0, 1\}\};$$

that is, each element of $\mathcal{B}_\infty$ is an infinitely long vector (ordered sequence) of 0's and 1's. Is $\mathcal{B}_\infty$ a countable set? What do your findings imply about infinity?

We now view the set of real numbers $\mathbb{R}$ as the collection of all decimal numbers of the form

$$I.d_1d_2d_3d_4 \ldots ,$$

where I is an integer and each digit d_n is an element of $\{0, 1, 2, \ldots , 9\}$.

7.8. Is $\mathbb{R}$ a countable set?

An infinite set S that is not countable is called *uncountable*.

The return of power sets

Recall that the power set $\mathcal{P}(S)$ of a set S is the collection of all subsets of S.

7.9. Prove and extend *or* disprove and salvage:

STATEMENT. *Let S be a set. Then the cardinality of* $\mathcal{P}(S)$ *is greater than the cardinality of S.*

7.10. Is there a set that has a greater cardinality than $|\mathbb{R}|$? If so, describe such a set. Is there one set larger than the one you have just described? If so, explain why and give an example of one element from the larger set. If not, explain why not.

7.11. Does there exist a "mother of all infinities?" That is, an infinite set whose cardinality is the biggest infinity?

7.12. Is $|\mathbb{R}|$ the "next" infinity after $|\mathbb{N}|$? That is, does there exist an infinite set S such that $|\mathbb{N}| < |S| < |\mathbb{R}|$?

7.13. A number α is called *algebraic* if there exists a nonzero polynomial $p(x)$ with integer coefficients such that $p(\alpha) = 0$. A number that is not algebraic is called *transcendental*. Show that transcendental numbers exist. (*Recall:* A *polynomial* is a function that can be expressed as

$$p(x) = c_N x^N + c_{N-1} x^{N-1} + \cdots + c_1 x + c_0$$

for some integer $N \geq 0$.)

A pair of paradoxes

We close our chapter on infinity with two paradoxes that lead to an important realization.

7.14. The Number Name Paradox. Let S be the set of all natural numbers that are describable in English words using no more than 50 characters (so $240 \in S$, because we can describe it as "two hundred forty," which requires fewer than 50 characters, and $25 \in S$, because we can describe it as "the integer that immediately follows twenty four," or more compactly as "five squared," or simply as "twenty five"). Assuming that we are allowed to use only the 27 standard characters (the 26 letters of the alphabet and the space character), show that there are only *finitely* many numbers contained in S. Now let the set $T = \mathbb{N} \setminus S$. Show that there are infinitely many elements in T and that T contains a smallest number. Finally let t be the smallest number contained in T. Prove that $t \in S$ and $t \notin S$. *Whoops.* What went wrong?

7.15. Bertrand Russell's Barber's Paradox. In a certain village, there is one male barber who shaves all those men, and only those men, who do not shave themselves. Does the barber shave himself? Show that the answer cannot be yes or no, and thus this question is a paradox. This barber paradox is related to another important insight due to Russell: Let *NoWay* denote the set of all sets that do not contain themselves as elements. Is $NoWay \in NoWay$? We now realize that we have yet to offer a precise mathematical definition of *set*. Given the above paradoxes, what is one critical property an actual set must possess?

Stepping back

Consider the following statement: *This statement is false or I will receive an A in my math class.* Could the statement be false? What can you conclude from this sentence? Can you modify the sentence to "prove" anything? What is wrong with the logic?

8

Recursively defined functions

The next generation

When a function is expressed as an algebraic expression (such as $f(x) = 3x^2 - x + 1$), we can easily analyze and study it. But some functions are described in more cryptic ways. Here we study functions that are defined not as an algebraic expression but inductively in terms of their previous values. We will discover a method for using this inductive definition to determine—in many cases—an honest-to-goodness algebraic expression for that function.

Moving ahead by looking back

By a *sequence,* we mean a function $a: \{0, 1, 2, \ldots\} \to \mathbb{R}$. By adopting the notation a_n for $a(n)$, we can give the values of the function $a(n)$ by listing the sequence $a_0, a_1, \ldots$. Some sequences are easy to express in terms of n—for example, the sequence 0, 1, 2, ... can be trivially expressed as the sequence $a_n = n$. In the sequence $p_n =$ (the nth prime number), however, listing the first few values—$p_1 = 2$, $p_2 = 3$, $p_3 = 5$, $p_4 = 7$, $p_5 = 11$—is easy, but finding an explicit general formula for p_n that depends only on the index n remains a complete mystery.

Here we examine an important class of sequences that, with respect to the difficulty in expressing the sequences in terms of only the index n, lies somewhere between the two previous examples. In these sequences, known as *recursively defined functions* or *recurrence sequences,* the term a_n depends on the previous terms of the sequence. We have seen these naturally occurring sequences many times. For example, consider the recursively defined sequences given by the following:

- $a_0 = 1$, $a_1 = 1$, $a_2 = 2$, and for all $n \geq 3$, $a_n = na_{n-1}$;
- $b_1 = 1$, and for $n \geq 2$, $b_n = b_{n-1} + n$;
- $c_0 = 1$, and for all $n \geq 1$, $c_n = 3c_{n-1}$; and
- $d_0 = 1$, $d_1 = 1$, and for all $n \geq 2$, $d_n = \frac{1}{2}(d_{n-1}+d_{n-2})$.

The *initial conditions* are the first few terms that are explicitly defined before the general recursive relation is given. For example, the initial conditions for the sequence a_n above are $a_0 = 1$, $a_1 = 1$, and $a_2 = 2$. Note that some sequences begin with the index $n = 1$ rather than $n = 0$.

8.1. Find *closed formulas* for the sequences a_n, b_n, c_n, and d_n defined in the previous paragraph; that is, find an explicit algebraic formula for each sequence ($a_n =$??; $b_n =$??, etc.) that involves only the index n and *not* the previous terms. Express your formulas as simply as possible (avoid using "...").

Here we develop a systematic method that allows us to find closed formulas for recursively defined sequences. By experimentation (also known as *trial and error*), we begin to develop some insights into this issue.

8.2. Consider the sequence defined by $a_1 = 1$ and for all $n \geq 2$, $a_n = 4a_{n-1} - 2$. Compute the first few terms in this sequence. Then by looking at these values, *guess* a closed formula for a_n (again, avoid "..." in your answer). Once you have your guess, *prove* that your conjecture is correct. (*Hint:* The answer might involve n as an exponent.)

Our main goal for this module is to answer the following question: How would we find a closed formula for a recurrence sequence such as the one above *without* guessing? The answer resides in a beautiful connection between recurrence sequences and formal power series.

Forgoing convergence—A formal look at power series

A *power series* is a function that has the basic shape $f(x) = \sum_{n=0}^{\infty} a_n x^n$, where the coefficients a_n are constants. In calculus, we were very concerned with questions of convergence and with finding the interval for x for which $f(x)$ exists. Here we disregard all issues of convergence and consider power series as formal objects; that is, we will never evaluate a formal power series at a particular x-value. Thus we can manipulate formal power series ... well, formally. For example, if $f(x) = \sum_{n=0}^{\infty} a_n x^n$ and $g(x) = \sum_{n=0}^{\infty} b_n x^n$, then we define their *sum* to be componentwise addition:

$$f(x)+g(x)=\sum_{n=0}^{\infty}(a_n+b_n)x^n$$

and their *product* to be a formal and elaborate version of "FOIL"ing from our binomial youth:

$$f(x)g(x)=\sum_{n=0}^{\infty}c_n x^n,$$

where $c_n = a_0 b_n + a_1 b_{n-1} + \dots + a_n b_0$. The best way to make this ultimate "FOIL"ing feat real is to attempt it—multiply termwise and combine like powers of x. For example, the product power series would begin

$$a_0 b_0 + (a_0 b_1 + a_1 b_0)x + (a_0 b_2 + a_1 b_1 + a_2 b_0)x^2 + \cdots.$$

Notice that in each monomial term $a_i b_j$, the two indices i and j sum to the exponent of x by which $a_i b_j$ is multiplied. Also observe that all such sums are present. In an important sense, this is the power of the formal power series—it allows us to group combinations of coefficients in systematic and useful ways by formal algebraic manipulations.

From calculus, recall the geometric series

$$\frac{1}{1-x}=1+x+x^2+x^3+\cdots \quad \left(\sum_{n=0}^{\infty}x^n, \text{ for short}\right),$$

which converges for all $|x| < 1$. Of course, we no longer care about convergence, so the previous identity could be proven *formally* by computing

$$\begin{aligned}&(1-x)(1+x+x^2+x^3+\cdots)\\&=1-x+x-x^2+x^2-x^3+\cdots+x^n-x^n+\cdots=1\end{aligned}$$

(verify this product for yourself) and then dividing both sides by $1-x$.

8.3. Find the formal power series for the following functions: $1/(1-(ax))$, where a is a constant; $x/(1+x)$; and $1/(1-x)^2$. Your answers should be of the form $\sum_{n=0}^{\infty} a_n x^n$.

8.4. Find the formal power series for $1/(4x^2 - 4x + 1)$.

We now connect formal power series with recurrence sequences in a most natural manner. Given a sequence $a_0, a_1, \ldots$, we define its *generating function* to be the formal power series given by

$$f(x) = \sum_{n=0}^{\infty} a_n x^n.$$

The power of moving from sequences to generating functions is illustrated in the following example. The method outlined in the next challenge can be applied to a wide class of recurrence sequences. Thus this challenge holds the key to unlocking all the challenges in this module.

8.5. Consider the recurrence sequence given by $a_0 = 1$ and for all $n \geq 1$, $a_n = 5a_{n-1}$. Let $f(x)$ be the generating function for this sequence. *Follow these steps:* Compute the formal power series $f(x) - 5xf(x)$, and use the recurrence relationship to show that most of the terms cancel. Divide your answer by $1 - 5x$ to produce an expression for $f(x)$ that does not involve any a_n terms. Express this new expression as a formal power series. This *new* formal power-series for $f(x)$ must equal the original one used to define $f(x)$ as the generating function. Set the corresponding coefficients of these two power-series representations of $f(x)$ equal to one another to discover a closed formula for a_n. *Very cool!*

Generating closed formulas through generating functions

The previous challenge illustrates an important principle: Given a recurrence sequence, we can formally manipulate its associated generating function in sympathy with the recurrence relation to produce a formal power series of the generating function that does not involve the original sequence. Equating the corresponding terms of these two equivalent power-series representations yields a closed formula for the recurrence sequence.

In Challenge 8.5, the critical step was the realization that we should consider the "sympathetic" function $f(x) - 5xf(x)$. Where did this come from? Notice that the recurrence relationship can be expressed as $a_n - 5a_{n-1} = 0$. This expression resembles, in some vague sense, the algebraic expression that led us to success. Notice the role of the factor of x in the second term: It allows us to align the coefficients so we can apply the recurrence relationship to dramatically simplify the power series for our algebraic expression. Revisit your work in Challenge 8.5 and notice the need for the extra factor of x. Once armed with these insights, can you extend them?

8.6. Consider the recurrence sequence given by $a_0 = 3$, $a_1 = -2$, and for all $n \geq 2$, $a_n = 2a_{n-1} - a_{n-2}$. Find a closed formula for a_n by studying the associated generating function.

8.7. Consider the recurrence sequence given by $a_0 = 1$, $a_1 = 4$, and for all $n \geq 2$, $a_n = 4a_{n-1} - 4a_{n-2}$. Find a closed formula for a_n by studying the associated generating function. (*Hint:* The answer to a previous challenge might be applicable here.)

8.8. Returning to the sequence defined by $a_1 = 1$ and for all $n \geq 2$, $a_n = 4a_{n-1} - 2$, for which a closed formula a_n was found by trial and error, use the theory of generating functions to provide further confirmation of the closed formula for a_n found in Challenge 8.2. (*Hint:* The technique of partial fractions used in integration theory may be useful here.)

8.9. Recall that we defined the *Fibonacci sequence* by $F_0 = 1$, $F_1 = 1$, and for all $n \geq 2$, $F_n = F_{n-1} + F_{n-2}$. Find a closed formula for the nth Fibonacci number. (*Hint:* Don't be worried if you need to pull out the ever-popular quadratic formula and get algebraically down and dirty.)

8.10. An application: When will the world end? The Towers of Hanoi is a puzzle consisting of three pegs and a collection of punctured disks of different diameters that can be placed on the pegs.

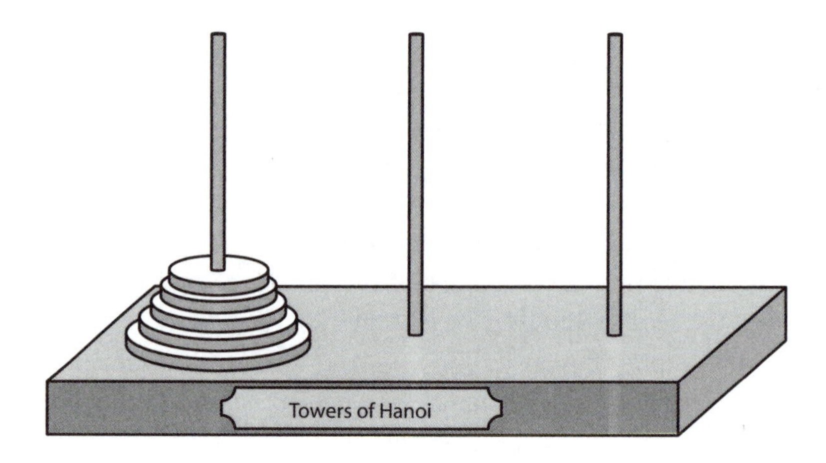

The puzzle begins with all disks on one peg in order of diameter, with the largest disk on the bottom.

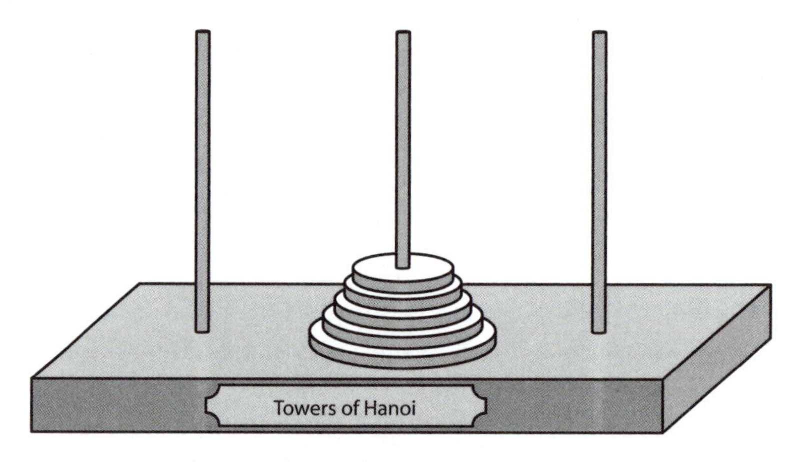

The object is to transfer all the disks to another peg so that they reside on this new peg in the original order given the following two rules: Only one disk can be moved to a peg at a time, and at no time can a larger disk be placed on top of a smaller disk.

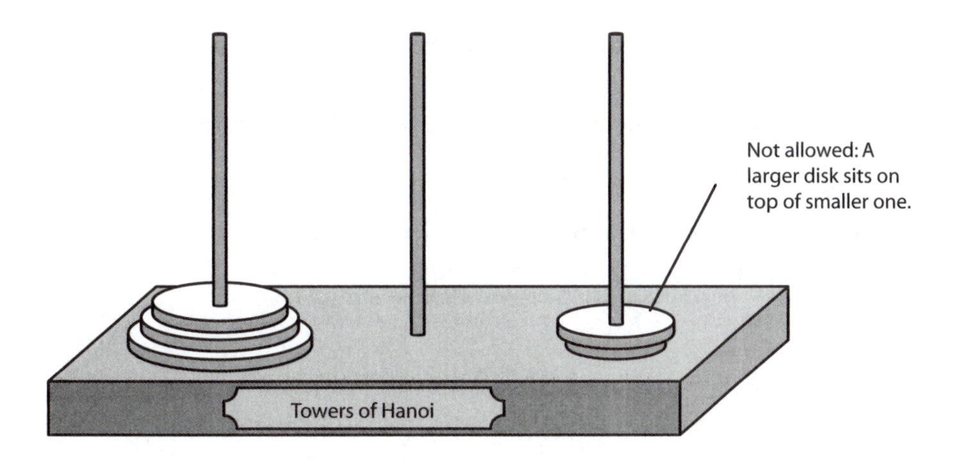

Describe a solution to the puzzle if there are n disks. Let h_n be the number of moves required to solve the puzzle with n disks. Find a recurrence relation for the sequence h_n. Solve the recurrence relation to find a closed formula for h_n. There is a legend that monks had a particularly impressive edition of this puzzle consisting of 64 gold disks and 3 diamond pins. They were able to move one disk per second. The legend is that the world would end once the monks completed their mission. Use your closed formula to predict when the world will end—a useful little piece of information as you plan your future.

Stepping back

Suppose that a recurrence sequence $\{s_n\}$ satisfies the recurrence relation

$$s_{n+1} = As_n + Bs_{n-1}$$

for some given constants A and B. Suppose also that we are given the values for s_0 and s_1. Another way to write the recurrence relation is $s_{n+1} - As_n - Bs_{n-1} = 0$. Inspired by this, we are led to consider the equation

$$x^2 - Ax - B = 0\,.$$

What thinking inspired this equation? Next find the two solutions to the quadratic equation—let's call them α_1 and α_2. Let's assume that α_1 and α_2 are distinct. Now show that there exist constants C_1 and C_2 such that for all $n \geq 0$,

$$s_n = C_1(\alpha_1)^n + C_2(\alpha_2)^n.$$

What determines the constants C_1 and C_2? How do your discoveries compare with your work in Challenge 8.9? *Neat stuff.*

9

Discrete thoughts of counting

Quantifying complicated quantities

There are many ways to "count" objects. We begin by considering insights that arise from rough estimation and a very simple principle, and then move to an exploration of *combinatorics*—an area of mathematics that involves precise counting. While counting might at first seem totally trivial, in many scenarios, it is a most challenging task. What is so hard? To count the elements of a collection accurately, we want to ensure that we count each element *and* that we count each element exactly once—so no element is not counted and no element is counted several times. Not so easy. The payoff of careful counting can be substantial, especially when we apply it to computing probabilities and determining the likelihood of future events.

Pigeons without a home

An important credo in mathematics is that subtle results often arise by studying simple phenomena very deeply. Here we see an extremely valuable tool in mathematics that appears so self-evident that proving it appears unnecessary.

9.1. Prove and extend *or* disprove and salvage:

Statement (The Pigeonhole Principle). *Let N be a natural number greater than* 1. *Suppose that N objects are to be placed in $N - 1$ boxes. Then at least two distinct objects would be placed in the same box.*

9.2. How many people are required to be in a room so that you are *certain* that two of them will share the same birthday (month and day)?

9.3. Prove and extend *or* disprove and salvage:

STATEMENT. *Let $a_0, a_1, a_2, \ldots, a_N$ be integers. Then there exist indices m and n such that N divides $a_n - a_m$.*

9.4. Prove and extend *or* disprove and salvage:

STATEMENT. *Let S be a set of four distinct points in the plane such that all four points are contained within or on the edges of a square having side length 2. Then there exist points $p_1, p_2 \in S$, with $p_1 \neq p_2$, satisfying $|p_2 - p_1| < \sqrt{2}$.*

(*Remark:* By $|p_2 - p_1|$, we mean the length of the line segment that connects the point p_1 to the point p_2.)

9.5. Prove and extend *or* disprove and salvage:

STATEMENT (The Happy Happy Thought). *In any group of 130,000 people, there must exist two individuals who share the same birthday (month and day) and will share the same death day (month and day).*

PERMUTATION = UP TO A REMINT

A *permutation* of a set of distinct objects is a linear ordering of the elements of the set. For example, 7, 3, 2, 9 is a permutation of the set {3, 2, 7, 9}, and UPTOAREMINT is a permutation of PERMUTATION.

9.6. List all the permutations of the set $\{A, E, T\}$.

9.7. Prove and extend *or* disprove and salvage:

STATEMENT. *Given n distinct objects, there are $n!$ permutations of the set; that is, there are $n!$ different ways of linearly ordering n different elements.*

(Recall that $n!$ denotes *n factorial*—that is, $n! = n(n-1)(n-2) \cdots (2)(1)$.)

9.8. How many permutations are there of the letters of the English alphabet taken four letters at a time? These are referred to as 4-*permutations*. Which of the following are 6-permutations of the alphabet:

BURGER

ROUNDS

CAMPUS

Find a formula for the number of r-permutations from a set containing n distinct objects.

9.9. How many ways can n sophomores and n seniors be seated at a round table if, for some sophomoric reason, no two sophomores can be seated next to each other? (*Remark:* Here we consider relative positions; that is, if the students are sitting around the table and they all get up and move to the right one seat, then that new seating configuration is the same as the first.)

A combination that unlocks counting

Suppose we are to create a committee to organize an all-campus party. There are 22 students who know how to throw a good party and are organized enough to pull it off. The committee must contain exactly 6 members. How many ways are there of selecting 6 students from the group of 22 potential party planners? This type of real-life question is a recurring one (not only within the context of committees, but also in many "game$" of chance). Here we develop the mathematical machinery to answer it. Party on!

9.10. Prove the following:

Theorem. *Given n objects, the number of ways to select m of them if the order in which the objects are selected does* ***not*** *matter is*

$$\frac{n!}{m!(n-m)!}.$$

The quantity given in the theorem is denoted $\binom{n}{m}$ and is read "n choose m." A subset of a set is known as a *combination*. Thus $\binom{n}{m}$ represents the number of m-combinations taken from a set of n objects.

9.11. How many different 5-card hands can be dealt from a standard 52-card deck of playing cards? How many of those 5-card hands are all hearts? How many are flushes (that is, all five cards are of the same suit)?

9.12. Verify the following: For all $r \le n$,

$$\binom{n}{r} = \binom{n}{n-r} \quad \text{and Pascal's famous identity} \quad \binom{n}{r-1} + \binom{n}{r} = \binom{n+1}{r}.$$

9.13. Pascal's identity leads to the well-known Pascal's triangle—an isosceles triangle of natural numbers whose first row is simply $\binom{0}{0}$, whose second row is $\binom{1}{0}\ \binom{1}{1}$, whose third row is $\binom{2}{0}\ \binom{2}{1}\ \binom{2}{2}$, and so forth. Draw the first seven rows of Pascal's triangle and show how the identity in Challenge 9.12 provides a visual method by which one row can be used to generate the next. Finally consider the sum of elements along various "diagonals" and see if you can discover any familiar patterns or sequences.

9.14. Prove the following:

THEOREM (The Binomial Theorem). *For any integer* $N \ge 0$,

$$(a+b)^N = \sum_{n=0}^{N} \binom{N}{n} a^n b^{N-n}.$$

In view of the Binomial Theorem, $\binom{n}{m}$ is often referred to as "the binomial coefficient n choose m."

9.15. Use the Binomial Theorem to *quickly* find a closed formula for the expression $\sum_{n=0}^{N} \binom{N}{n}$. Once you have your answer, find a connection with the power set of the set $\{1, 2, \ldots, N\}$. Apply this new insight to give a combinatorial proof of your newly discovered identity that does *not* require the Binomial Theorem.

Stepping back

Consider Pascal's triangle but with the entries expressed mod 2 (that is, 0 if the entry is even and 1 if the entry is odd). What famous image comes into focus as we consider more and more levels of the triangle? Can you give a reason why we see this beautiful phenomenon?

10

Quantifying uncertainty with probability

A likely story?

We do not know what will happen in the future. Using mathematics, however, we can place a numerical value to the likelihood that a certain event will occur. The higher the value, the more probable it is that the event will occur. This quantification of likelihood is known as the *theory of probability.* Here we explore the basics of this beautiful and important area and, along the way, make some counterintuitive realizations.

What are the chances?

By an *experiment* $\mathcal{E}$ we mean some activity for which there are only finitely many possible outcomes—all known in advance, while the actual outcome is uncertain. Thus it follows that an experiment must have more than one possible outcome. For example, $\mathcal{E}$ might be the act of flipping a coin three times.

The *sample space* $\mathcal{S} = \mathcal{S}(\mathcal{E})$ of an experiment $\mathcal{E}$ is the set of all possible outcomes. An *event* A is a subset of the sample space. Thus if $\mathcal{E}$ is the act of flipping a coin three times and we write H for heads and T for tails, then

$$\mathcal{S}(\mathcal{E}) = \{\mathsf{HHH, HHT, HTH, HTT, THH, THT, TTH, TTT}\},$$

and the subset $A = \{\mathsf{HHH, TTT}\}$ is an example of an event—specifically, the event in which all three flips land the same way.

In view of our studies into power sets, we can immediately conclude that if the sample space of a certain experiment has cardinality n, then there are 2^n distinct possible events.

For the remainder of this discussion, we will assume that $\mathcal{E}$ is an experiment having a sample space $\mathcal{S}$ in which any outcome is just as likely to occur as any other. We define a *probability function* $P : \mathcal{P}(\mathcal{S}) \to \mathbb{R}$ by

$$P(A) = \frac{|A|}{|\mathcal{S}|},$$

where the event $A \in \mathcal{P}(\mathcal{S})$. We refer to $P(A)$ as the *probability that the event A occurs.*

10.1. Prove and extend *or* disprove and salvage:

STATEMENT. *If A and B are events in $\mathcal{P}(\mathcal{S})$, then $P(A \cup B) = P(A) + P(B)$.*

10.2. Verify the following properties of a probability function:

- $P(\mathcal{S}) = 1$
- $\sum_{s \in \mathcal{S}} P(\{s\}) = 1$
- For $A \in \mathcal{P}(\mathcal{S})$, we define the *complement of A* as $\bar{A} = \mathcal{S} \setminus A$.
  $P(\bar{A}) = 1 - P(A)$
- $P(\emptyset) = 0$

10.3. Verify the following properties of a probability function:

- For any $A \in \mathcal{P}(\mathcal{S})$, $P(A) \geq 0$.
- If $A \subseteq B \subseteq \mathcal{S}$, then $P(A) \leq P(B)$.
- For any $A \in \mathcal{P}(\mathcal{S})$, $0 \leq P(A) \leq 1$.

10.4. Suppose someone has randomly selected two numbers from the set of the first one million natural numbers and used them to make a fraction. Reduce the fraction to its lowest terms. Is there a 0.5 probability that both the numerator and the denominator are odd numbers? Justify your answer.

A coin toss?

Here we consider likelihoods of repeated experiments. For example, if we were to flip a coin twice, then there would be a total of four possible, distinct outcomes: first heads, then heads; first heads, then tails; first tails, then heads; and finally first tails, then tails. We are beginning to see that counting (a.k.a. combinatorics) plays a central role in computing probability.

10.5. A coin is called *fair* if it is equally likely to land on its head or tail when flipped. Suppose a fair coin is flipped ten times. What is the probability of seeing exactly five heads? What is the probability of seeing exactly five heads all *in a row*? What is the probability of seeing at least one head?

10.6. Three different fair coins are flipped. What is the probability that they either all land heads up or all land heads down? What is your response to the following reasoning? *When we toss three coins, we know for a fact that two of the coins will be the same, thus we only have to get the third coin to match. Thus the probability is* $\frac{1}{2}$.

What's the deal?

We close our discussion of likelihoods with a look at how likely certain seemingly rare events truly are.

10.7. Using a standard 52-card deck of playing cards, select a card at random, record it, and then place it back in the deck. Shuffle the cards, select another card at random, and record it. Put it back in the deck. Repeat this process. How many cards did you draw before you selected a card that already appears on your list (that is, you see a match)? Attempt this experiment a few times. Now compute the probability of choosing ten times and seeing ten *different* cards. Using your answers, find the probability that after ten times, you have at least one match. Are you surprised by your answer?

10.8. Suppose someone shows you three cards. One is red on both sides, another is blue on both sides, and the last is red on one side and blue on the other. The cards are shuffled and flipped at random, and you are then shown one side of one card. You see red. What is the probability that the other side is blue? Explain your answer.

10.9. Suppose you are dealt five cards from a standard 52-card deck of playing cards. What is the probability of being dealt four aces? What is the probability of being dealt a flush? What is the probability of being dealt a full house (a full house is three of a kind and two of another kind—for example, 4 ♣, 4 ♢, 4 ♡, 7 ♢, 7 ♠ dealt in any order)?

10.10. Assume that a person's birthday is equally likely among the 366 days in a year. Suppose that 40 people are at a party. What is the probability that at least two of them share the same birthday? What is the smallest number of people needed so that the probability that at least two of them share the same birthday is greater than $\frac{1}{2}$?

Stepping back

Consider the following game of chance: You pay a certain amount of money to play. Then you flip a fair coin. If you see tails, you flip again; the game continues until you see a head, at which time the game ends. If you see a head on the first flip, you receive $2. If you see a head on the second flip, you receive $4; see a head on the third flip, you are awarded $8; and so forth. The payoff is doubled every time the coin is flipped. What is the expected payoff of this game? How much would you pay to play this game? Suppose you paid $1,000 to play. What is the probability that you will make a profit? Why is this game a paradoxical situation given the expected payoff? This famous game is known as *The St. Petersburg Paradox*.

11

The subtle art of connecting the dots

Edging up to graphs

Suppose we wish to paint the center line on a grid of streets and we cannot backtrack. What is the path to take? Suppose we wish to deliver mail to an area of homes. What is the most efficient route? Suppose we wish to design a network. How can we find the most cost-effective way to proceed? The answers to these questions can be found through the ideas and techniques of the modern study of graph theory. Here we explore the basic notions and results of this important area of discrete mathematics.

Bridging the graph

It was a cold wintery day in eighteenth-century Königsberg, the capital of East Prussia. People bundled up to keep warm while traveling up and down the seven bridges that connected the four landmasses separated by the Pregel River.

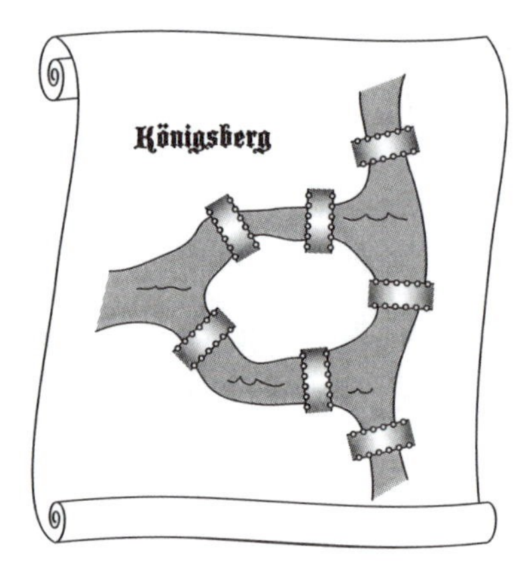

Given that there was no cable TV back then to offer mind-numbing entertainment, the locals amused themselves with the following mind-numbing conundrum: Is it possible to take a stroll in which all bridges are traversed exactly once *and* in which the walk ends at the same place it began? This question, known as the *Königsberg Bridge Problem*, was answered by Euler in 1736. His methods and ideas led to the modern area of mathematics known as *graph theory*.

Let V be a finite, nonempty set, and let E be a set consisting of subsets of V of the form $\{a, b\}$ such that a and b are in V. The pair (V, E) is called a *graph* (*on* V), and we call V the set of *vertices* and E the set of *edges*.

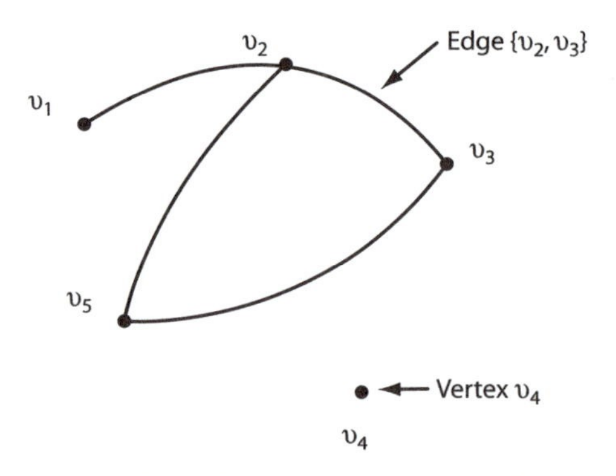

A sample graph with 5 vertices and 4 edges

We will write $G = (V, E)$. We will allow E to contain multiple $\{a, b\}$'s (thus we will slightly loosen our interpretation of "set" and "subset"). For example, if $V = \{a, b\}$ is the set of vertices and we have edges $E = \{\{a, b\}\}$, then the graph $G = (V, E)$ is different from $G' = (V, E')$, where E' is the set of edges $E' = \{\{a, b\}, \{a, b\}\}$.

Two different graphs having the same set of vertices

We will consider the edges $\{a, b\}$ and $\{b, a\}$ to be the same. (In some cases, it is beneficial to distinguish $\{a, b\}$ and $\{b, a\}$. If we do, the graph is called a *directed graph.*)

We say that a and b are the *endpoints* of the edge $e = \{a, b\}$, that an edge e *joins* its endpoints and is *incident* on a and b, and that a and b are *adjacent.* An edge of the form $\{a, a\}$ is called a *loop* at a. In this case, a is adjacent to itself. If v is a vertex, then we define the *degree* of v, denoted $\deg(v)$, to equal the number of edges incident on v, where a loop at v is counted twice. The *total degree* of G is the sum of the degrees of the vertices of G.

11.1. Prove and extend *or* disprove and salvage:

STATEMENT. *The total degree of a graph is even.*

11.2. Prove and extend *or* disprove and salvage:

STATEMENT. *Given any graph, the number of odd degree vertices is even.*

11.3. Determine whether the following graphs can exist. If the graph can exist, draw the graph; if not, state why not.

- $V = \{a, b, c, d\}$ with $\deg(a) = 2$, $\deg(b) = 1$, $\deg(c) = 5$, $\deg(d) = 0$
- $V = \{a, b, c, d\}$ with $\deg(a) = 1$, $\deg(b) = 4$, $\deg(c) = 2$, $\deg(d) = 2$
- $V = \{a, b, c, d\}$ with $\deg(a) = 1$, $\deg(b) = 3$, $\deg(c) = 2$, $\deg(d) = 5$

Making the circuit

A *walk* from a to b is a finite sequence of adjacent vertices and edges of G of the form $a, e_1, v_1, e_2, \dots, v_k, e_{k+1}, b$, where the v_i's are vertices of G and for each i, e_i is the edge $\{v_{i-1}, v_i\}$ (we declare $v_0 = a$ and $v_{k+1} = b$). A *path* from

a to b is a walk that contains no repeated edges. In other words, a path is a walk for which the e_i's above are distinct. A *simple path* from a to b is a path that contains no repeated vertices. A *closed walk* is a walk that starts and ends at the same vertex.

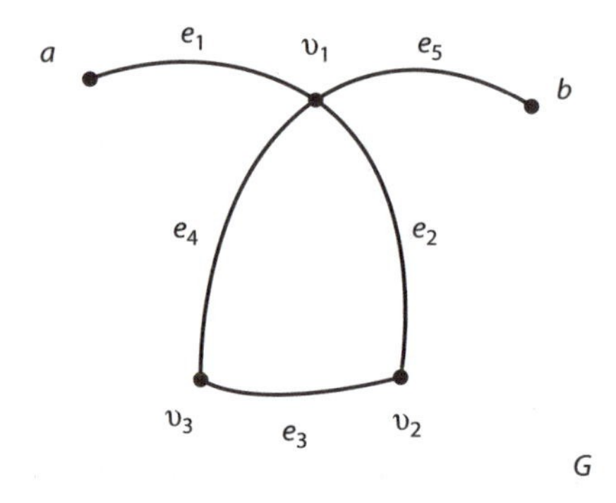

The walks a, e_1, v_1, e_5, b and $a, e_1, v_1, e_2, v_2, e_3, v_3, e_4, v_1, e_5, b$ are both paths from a to b. The first is a simple path, whereas the second is not simple. The walk $v_1, e_2, v_2, e_3, v_3, e_4, v_1$ is a closed walk.

A *circuit* is a closed walk that does not contain a repeated edge. A *simple circuit* is a circuit that does not have any other repeated vertex except the first and last. The graph G is *connected* if for each pair of distinct vertices a and b, there exists a walk from a to b. An *Euler circuit* for G is a circuit that contains every vertex and every edge of G. An *Euler path* from vertices a to b is a walk that starts at a, ends at b, and passes through every vertex of G at least once and traverses every edge of G exactly once.

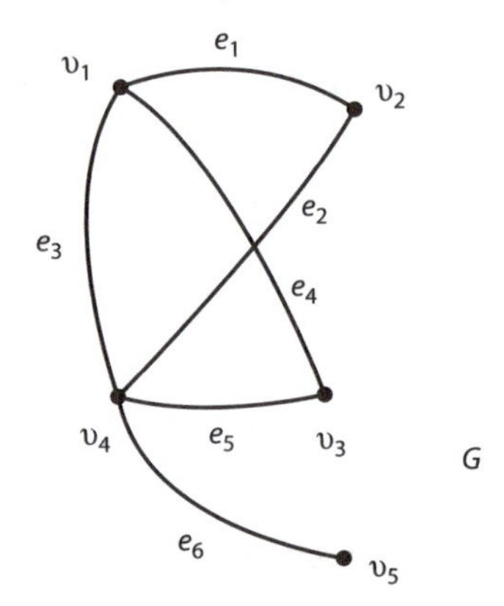

The walk $v_1, e_1, v_2, e_2, v_4, e_3, v_1, e_4, v_3, e_5, v_4, e_6, v_5$ is an Euler path from v_1 to v_5. Is there an Euler path from v_3 to v_5?

11.4. Prove and extend *or* disprove and salvage:

STATEMENT. *If G is connected and a and b are distinct vertices of G, then there exists a simple path from a to b.*

11.5. Prove and extend *or* disprove and salvage:

STATEMENT. *If a and b are vertices of a circuit on G and an edge e is removed from the circuit (while e's incident vertices remain), then there exists a path from a to b in this modified graph* $G \setminus \{e\}$.

11.6. Prove and extend *or* disprove and salvage:

STATEMENT. *If G is connected and contains a circuit C, then an edge of C can be removed without disconnecting G.*

11.7. Prove and extend *or* disprove and salvage:

STATEMENT. *A graph G has an Euler circuit if and only if it is connected and every vertex has even degree.*

11.8. Prove and extend *or* disprove and salvage:

STATEMENT. *If a vertex of G has odd degree, then G cannot contain an Euler circuit.*

11.9. Prove and extend *or* disprove and salvage (with another "if and only if" statement):

STATEMENT. *The graph G has an Euler path if and only if G has zero or two vertices of odd degree.*

11.10. Solve the Königsberg Bridge Problem.

Branching out to trees

A graph is *circuit-free* if it contains no circuits. A graph is called a *tree* if it is connected and circuit-free.

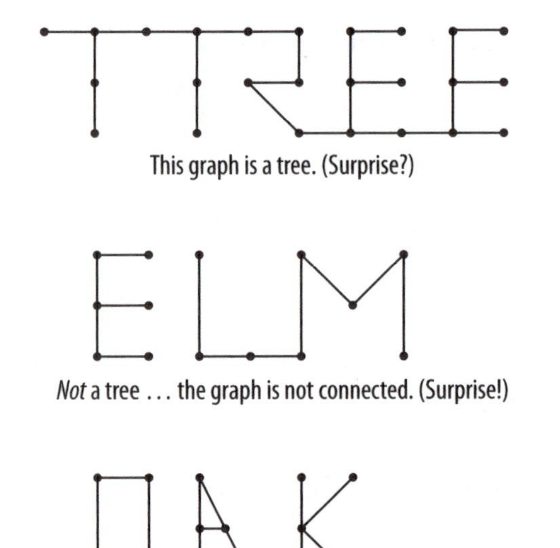

This graph is a tree. (Surprise?)

Not a tree . . . the graph is not connected. (Surprise!)

Not a tree . . . the graph has two circuits. (Surprise again!)

11.11. Prove and extend *or* disprove and salvage:

STATEMENT. *Any tree that has more than one vertex has a vertex of degree* 1.

11.12. Prove and extend *or* disprove and salvage (with another "if and only if" statement):

STATEMENT. *A graph G with n vertices is a tree if and only if G is connected and has n edges.*

11.13. Prove and extend *or* disprove and salvage:

STATEMENT. *If a and b are distinct vertices of a tree G, then there is a unique path between these vertices.*

Completely connected graphs

For a positive integer n, we define the *complete graph on n vertices*, written as $\mathcal{K}_n$, to be the graph containing no loops and having n vertices, each pair of which are adjacent and connected by a unique edge.

11.14. Draw the following graphs: $\mathcal{K}_3$, $\mathcal{K}_4$, $\mathcal{K}_5$.

11.15. Prove and extend *or* disprove and salvage:

STATEMENT. *The graph $\mathcal{K}_n$ contains $[n^2/2]$ edges, where $[x]$ denotes the integer part of x.*

Stepping back

Show that for any graph $G = (V, E)$ having no loops or multiple edges, the following relationship between the cardinality of V and E holds:

$$|E| \le \frac{|V|(|V|-1)}{2}.$$

12

Just plane graphs

Drawing without being cross

Here we delve further into the theory of graphs as we focus on which graphs, in some sense, are naturally at home in the plane. These graphs exhibit some graceful structure. As we will see in Module 13, the theory developed here can be applied to generate new insights into geometry.

Drawing without crossing

A graph G is *planar* if it can be drawn in the plane such that the edges of G do not intersect except possibly at vertices of G.

12.1. Prove and extend *or* disprove and salvage:

STATEMENT. *Let G be a tree. Then G is a planar graph.*

We remark that a planar graph divides the plane into a finite number of regions, including the unbounded region "outside" G.

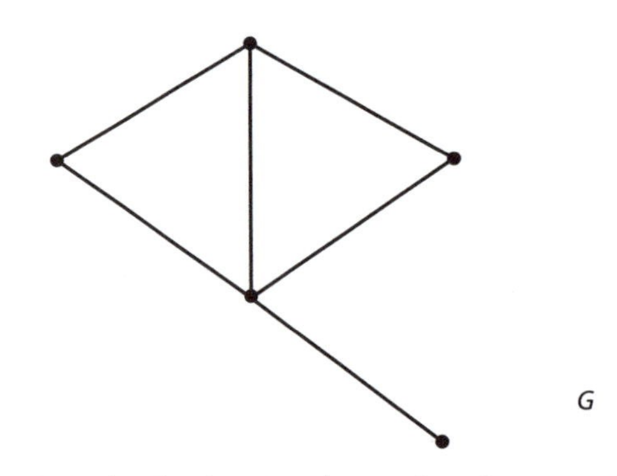

This graph divides the plane into three regions. Do you see them all?

Suppose that G is a connected and planar graph with V vertices and E edges. If G divides the plane into F regions, then we define the *Euler characteristic of* G to equal the quantity $V - E + F$.

12.2. Prove and extend *or* disprove and salvage:

STATEMENT. *If* G *is a connected, planar graph, then the Euler characteristic of* G *is* 2.

For a positive integer n, we recall that $\mathcal{K}_n$ denotes the complete graph on n vertices. For positive integers m and n, we now define the *complete bipartite graph* $\mathcal{K}_{m,n}$ to be a graph having $m + n$ vertices, in which each of the *first* m vertices is adjacent to each of the *last* n vertices.

12.3. Draw the following graphs:

- $\mathcal{K}_3$
- $\mathcal{K}_4$
- $\mathcal{K}_5$
- $\mathcal{K}_{2,3}$
- $\mathcal{K}_{3,3}$
- $\mathcal{K}_{2,4}$

Which of these appear to be planar graphs?

12.4. Prove and extend *or* disprove and salvage:

STATEMENT. *Let G be a connected, planar graph having V vertices and E edges. If $V \geq 3$, then $E \leq 2V - 6$.*

12.5. Prove that the graph $\mathcal{K}_5$ is not a planar graph.

12.6. Prove and extend *or* disprove and salvage:

STATEMENT. *Let G be a connected, planar, complete bipartite graph having V vertices and E edges. If $V \geq 3$, then $E \leq 2V - 4$.*

12.7. Prove that the graph $\mathcal{K}_{3,3}$ is not a planar graph. Then resolve the *Gas, Water, Electric Puzzle:* There are three houses, and each has a gas line (running from the gas company), a water line (running from the water company), and an electric line (you guessed it, running from the electric company). Prove that it is impossible to draw all the gas, water, and electric lines in the plane without having some of the lines cross one another.

Seemingly similar graphs

Let $G = (V, E)$ and $G' = (V', E')$ be two graphs having no loops or multiple edges. We say that G and G' are *isomorphic graphs* if there exists a one-to-one, onto function $f: V \to V'$ such that for every $a, b \in V$, $\{a, b\} \in E$ if and only if $\{f(a), f(b)\} \in E'$. That is, two graphs are isomorphic if, once we change the names of the vertices, the graphs are identical.

For any natural number n, we write W_n for the graph whose set of vertices is given by $V = \{0, 1, 2, \ldots, n\}$ having edges $E = \{\{0, 1\}, \{0, 2\}, \ldots, \{0, n\}, \{1, 2\}, \{2, 3\}, \ldots, \{n-1, n\}\}$. W_n is called the *n-spoked wheel graph* (Why?). For example, W_3 looks like

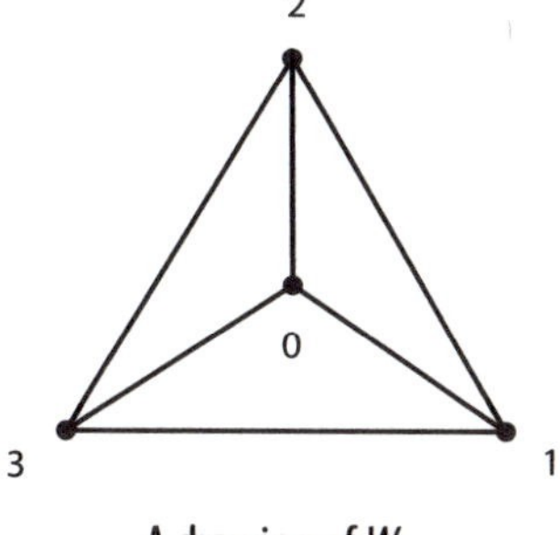

A drawing of W_3.

It is easy to verify that W_3 is isomorphic to $\mathcal{K}_3$.

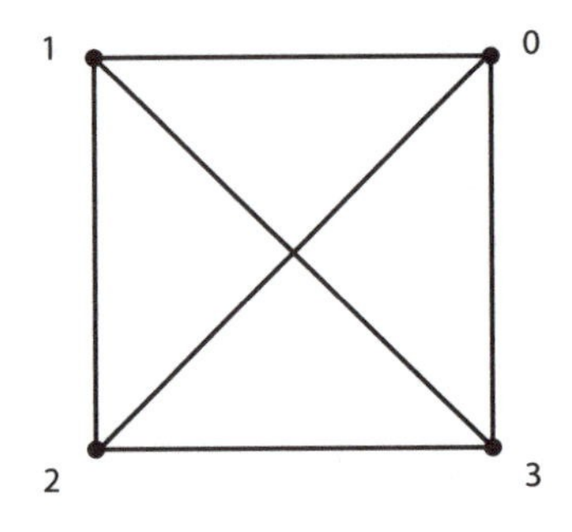

12.8. Prove and extend *or* disprove and salvage:

STATEMENT. *If W_n is isomorphic to W_m, then $n = m$.*

12.9. Prove and extend *or* disprove and salvage:

STATEMENT. *The following two graphs are isomorphic.*

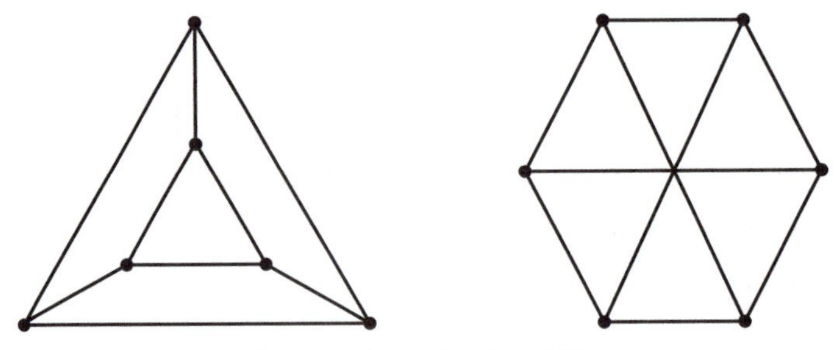

A not-so-planar drawing of $\mathcal{K}_4$

If a graph is not planar, we can locate particular "pieces" that make it nonplanar. A *subgraph* of a graph G is a graph with vertices and edges contained in G. For example, in the left graph in the previous figure, the three vertices and edges in the center form a subgraph that happens to be isomorphic to $\mathcal{K}_3$.

12.10. Prove and extend *or* disprove and salvage:

STATEMENT. *If G has a subgraph isomorphic to $\mathcal{K}_{3,3}$, then G is not planar.*

12.11. Prove and extend *or* disprove and salvage:

STATEMENT. *If G is not planar, then it contains a subgraph isomorphic to $\mathcal{K}_5$ or $\mathcal{K}_{3,3}$.*

While the statement in Challenge 12.11 is challenging, it can be generalized. This important extension requires a more general relation between graphs than the notion of isomorphism. We say that a graph *G* is *homeomorphic* to a graph *H* if *G* can be obtained from *H* by adding vertices to the edges of *H*. By necessity, these vertices must be of degree 2.

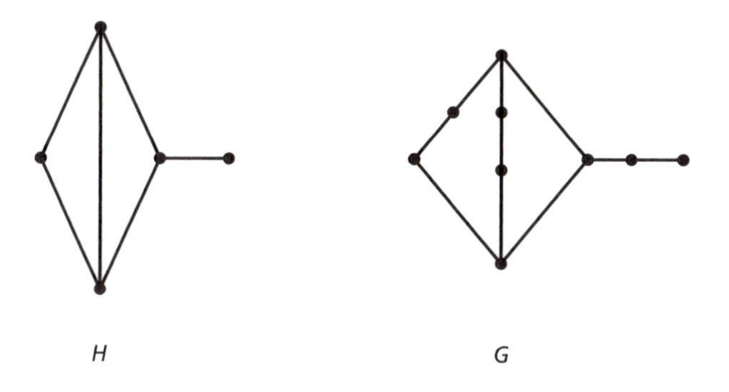

12.12. Prove and extend *or* disprove and salvage:

STATEMENT. *The graph below contains a subgraph homeomorphic to $\mathcal{K}_4$.*

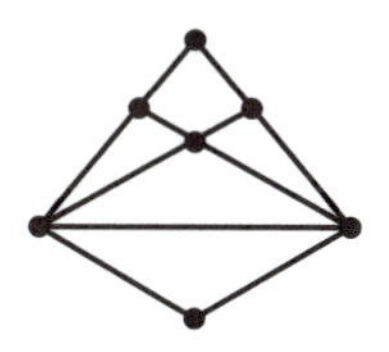

12.13. Prove that the graph below, known as the *Peterson graph,* is not planar.

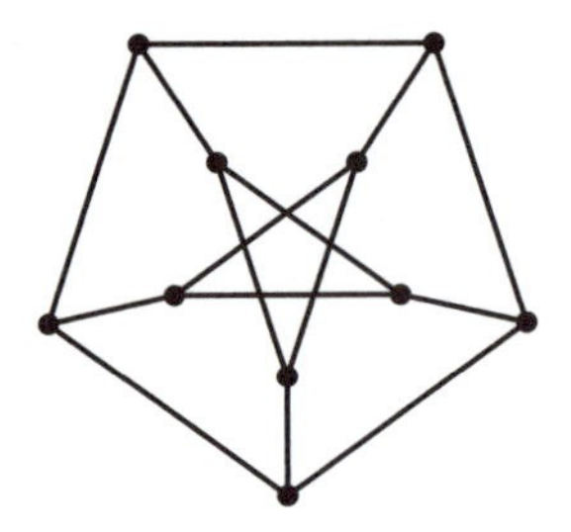

The notion of homeomorphic graphs allows us to appreciate the importance of the graphs $\mathcal{K}_5$ and $\mathcal{K}_{3,3}$. In some sense, these graphs are the building blocks of all nonplanar graphs. Specifically, we state the following deep theorem (without proof).

THEOREM (Kuratowski's Theorem). *A graph is planar if and only if it contains no subgraph homeomorphic to* $\mathcal{K}_5$ *or* $\mathcal{K}_{3,3}$.

Sensing symmetry

We can say that a connected planar graph G has a *symmetric planar drawing* if

- all its vertices have the same degree,
- *G* can be drawn in the plane so that its edges intersect only at the vertices of *G*,
- there is a planar drawing of *G* such that each region in the plane bounded by the graph is bounded by the same number of edges.

12.14. Consider the following three graphs and determine which have a symmetric planar drawing and which do not. Justify your answers.

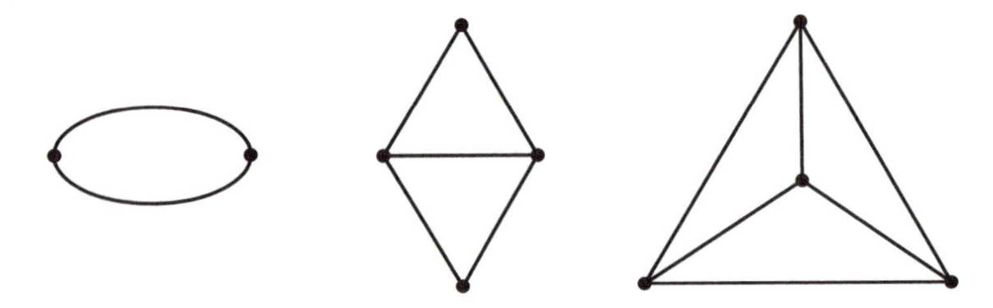

We say a graph is a *regular planar graph* if it has a symmetric planar drawing in which each vertex has degree at least 3, each region in the plane bounded by the graph has at least three edges, and each edge is a boundary for two distinct regions. Notice, for example, that the right-most graph in the figure above is an example of a regular planar graph.

12.15. Prove and extend *or* disprove and salvage:

STATEMENT. *There are nine regular planar graphs.*

Stepping back

Show that if G is a regular planar graph having V vertices and E edges, then for each vertex v in G, $\deg(v) = 2E/V$. Can you extend this result to a larger class of graphs?

13

Visible and invisible universes

Geometric vignettes

Geometry is one of the oldest areas of mathematics. It has inspired people throughout the ages to appreciate beauty and symmetry and discover hidden visual patterns. The power of abstract mathematical thinking is that it can take us to geometrical worlds that can only be viewed with the mind's eye. Here we briefly explore three areas within the geometric realm.

Sizing up symmetry

The most symmetric object in the plane is the perfectly smooth circle—the boundary of a disk. From every angle, it looks the same.

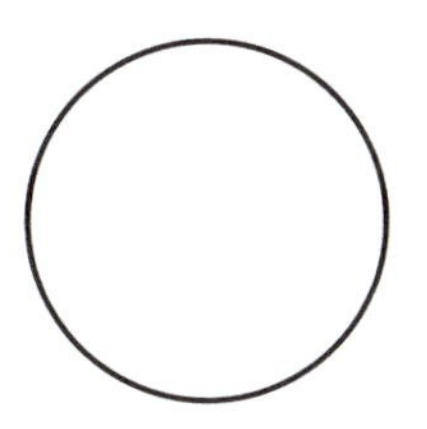

If we want to capture its beautiful symmetry with objects that have *flat* sides, we turn to the piecewise linear world of polygons. A *polygon* is a closed loop (in the plane) made from a finite number of line segments and with the property that the loop does not cross itself. For example, the first curve in the following figure is not a polygon because it does not form a loop. The second curve crosses itself and thus is also not a polygon. The third curve, however, is an example of a polygon—one that somewhat resembles a person trying to get down and "boogie."

A *regular polygon* is a polygon with the properties that any two sides have the same length and the measures of any two interior angles formed by adjacent edges are equal. For example, a square is a regular polygon. Given that these objects have a similar look when viewed from a variety of vantage points, the regular polygons are the most symmetric closed curves (in the plane) made with straight edges.

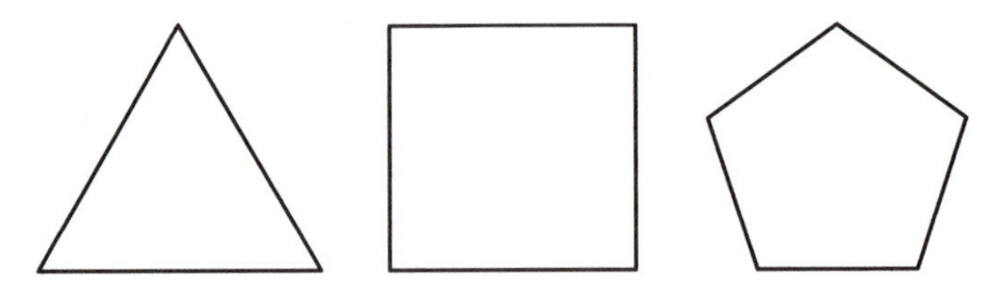

13.1. How many regular polygons are there? Justify your answer. Can you sketch them all?

If we now move to three-dimensional space, the issue becomes a bit more interesting. We see that the sphere (the surface of a ball) is the most perfectly symmetric object.

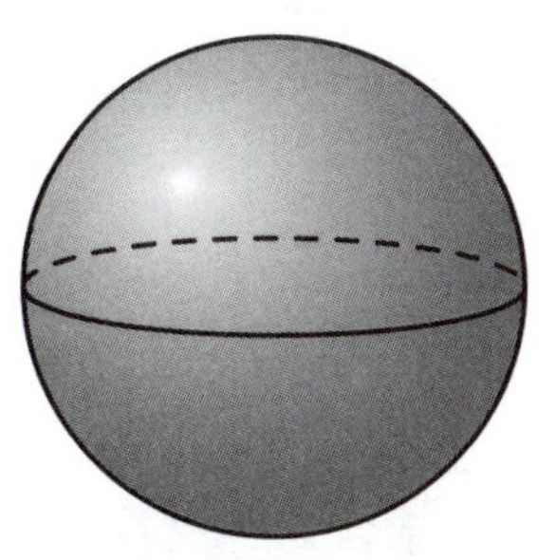

If we restrict ourselves to the surfaces of more basic solids with flat faces, then we consider the three-dimensional analogue of a polygon. A *polyhedron* is the surface of a three-dimensional solid whose boundary is a collection of polygons joined along their edges.

We now restrict our consideration to the most symmetric polyhedra. A polyhedron is called a *regular polyhedron* if all its faces are the same regular polygon and the number of edges emanating out from any one vertex is the same as from any other vertex. For example, a cube is a regular polyhedron, because every face is a square of the same size and there are three edges emanating out of any vertex.

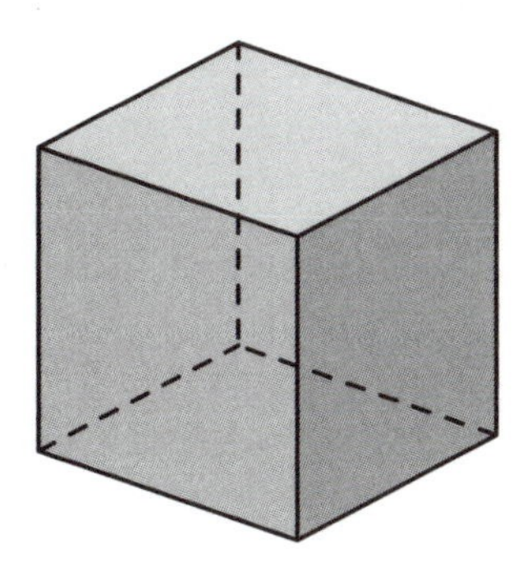

The regular polyhedra are the most symmetric flat-faced solids in space. Regular polyhedra, together with their interiors, are referred to as *regular solids.*

13.2. How many regular polyhedra are there? Justify your answer. Can you sketch them all? (*Hint:* Imagine a regular polyhedron whose edge-skeleton has been drawn on an extraordinarily elastic balloon, with the balloon's airhole in the center of one of the faces.

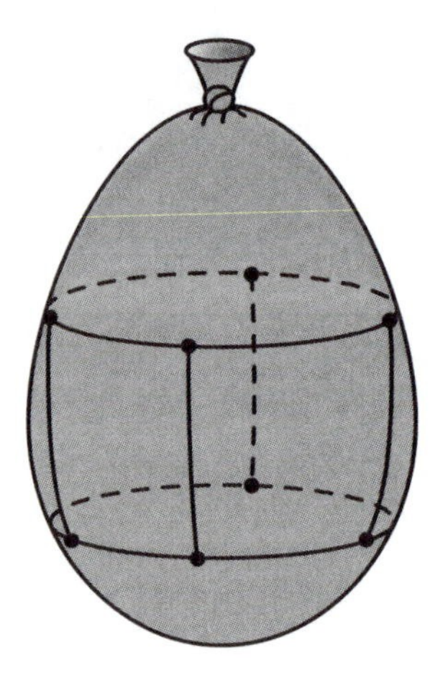

Now imagine untying the knot and stretching that airhole wide open so the balloon becomes a huge flat rubber disk that can be placed on the plane. Now consider the skeleton in the plane and apply one of the challenges from Module 12.)

13.3. For each regular polyhedron, devise a method for counting its number of faces, edges, and vertices. Make a chart, with the regular polyhedra listed down the first column and the number of faces, edges, and vertices for each listed in the respective rows. Fill in the chart and report all the patterns you discover.

Regular solid	Number of faces	Number of edges	Number of vertices
⋮	⋮	⋮	⋮

Let P be a regular polyhedron. The *dual polyhedron* of P is the polyhedron formed by first placing its vertices in the centers of the faces of P and then defining its edges to be those line segments that connect one vertex v to all those vertices that lie on faces of P that are adjacent to (that is, share a common edge with) the face of P containing v. We denote the dual of P as Dual(P).

13.4. Prove that the dual of a regular polyhedron is also a regular polyhedron. What does the information from the chart you made in Challenge 13.3 reveal with respect to duality?

13.5. Prove and extend *or* disprove and salvage:

STATEMENT. *Let P be a regular polyhedron. Then* Dual(Dual(P)) = P.

Keeping an artful eye on art galleries

Suppose an art gallery is housed within a polygon and contains no interior walls. The following are some examples of floor plans for such an art gallery.

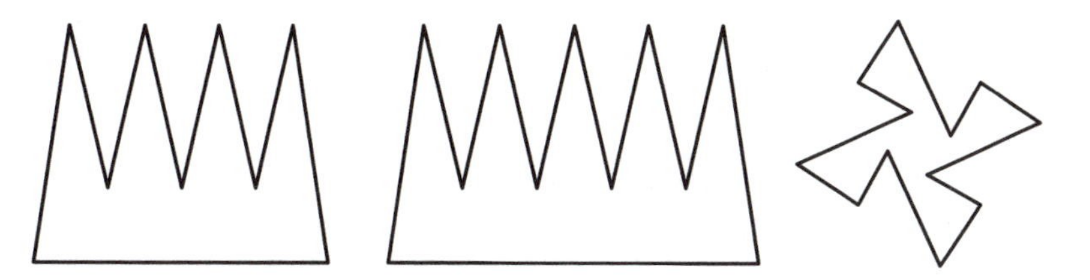

Because we wish to keep a watchful eye on all points within the gallery, we place cameras at some of the vertices of the polygon. Each camera is equipped with a fancy fish-eye lens and can see everything within its line of sight between the two adjacent edges flanking the vertex.

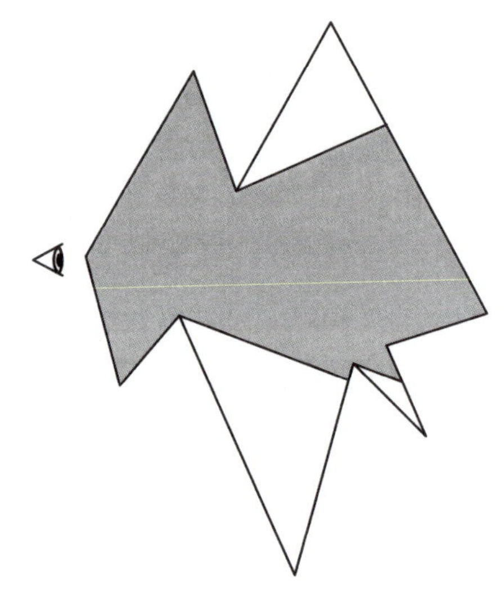

Shaded region = area that can be viewed from the camera eye

The fundamental question that now arises is: What is the smallest number of cameras required to view the entire interior of the polygon gallery? Phrased differently: Given a polygon with n vertices, what are the fewest vertices that allow the entire interior of the polygon to be viewed from those vertices? Our goal here is to answer this art gallery visibility question.

13.6. For each of the three polygonal art galleries above, find the fewest vertices required to view the interior. Which vertices would you select? Make a conjecture as to the fewest number of vertices required if the polygon has n vertices. Test your conjecture with several additional examples. Does your conjecture hold even with the most "spiky," comb-like polygonal galleries?

13.7. Prove and extend *or* disprove and salvage:

STATEMENT. *Let P be a regular polygon in the plane and let v be a vertex of P. Then there exists a line segment that lies completely in the interior of P and that connects v to another vertex of P that is not adjacent to v.*

A line segment as described in the previous statement is called a *spanning arc.*

13.8. Prove and extend *or* disprove and salvage:

STATEMENT. *Given any polygon P in the plane, there exists a unique manner of partitioning the interior of P into triangles so that the vertices of P are precisely the vertices of the triangles.*

A partition as described in the previous statement is called a *triangulation.*

13.9. Prove and extend *or* disprove and salvage:

STATEMENT. *Given any polygon P in the plane that has been triangulated, there exists a way of painting the vertices using three colors such that each triangle is painted with all three colors.*

13.10. Prove and extend *or* disprove and salvage:

STATEMENT. *Given any polygon P having n vertices in the plane, there exist* $[n/3]$ *vertices from which every point from the interior of P can be viewed.*

(Recall that $[x]$ denotes the greatest integer less than or equal to x.)

Up, up, and away—The fourth dimension

There are many different ways to define what we mean by dimension. Let's informally agree that the *dimension* of a space is the fewest pieces of information required to precisely locate an arbitrary point in that space given a set of coordinate axes.

13.11. Prove that $\mathbb{R}^N$ has dimension N.

We define a *0-dimensional cube* to be a point. If we imagine inking up that 0-dimensional cube and dragging it 1 unit in a new direction so that the ink leaves behind a trail of the dragging, then we would produce a *1-dimensional cube* (a.k.a. a line segment). If we now ink up the entire line segment and drag it 1 unit in a new, perpendicular direction, the inky residue would create a *2-dimensional cube* (better known as a square).

0-, 1-, 2-dimensional cubes

13.12. Continue the cube-construction procedure outlined above and offer a means of describing 3-, 4-, and 5-dimensional cubes. Provide sketches of 0-, 1-, 2-, 3-, 4-, and 5-dimensional cubes.

13.13. Let $C(n)$ be an n-dimensional cube. Fill in the following chart. How many patterns can you find in your table? Can you find general formulas for extending the table indefinitely for any n?

Dimension n	Number of vertices	Number of edges	Number of 2-D faces	Number of 3-D "faces"	Number of 4-D "faces"
0					
1					
2					
3					
4					
5					
n					

Searching for patterns in n-dimensional cubes

13.14. Suppose that $C(n)$ is an n-dimensional cube of edge length equal to 1. We say that a point $p \in C(n)$ is *near the boundary of the cube* if it is not contained in the concentric cube having edge length equal to 1/2. Show that as n increases, the probability that a randomly selected point in $\mathcal{C}(n)$ is near the boundary of the cube approaches 1.

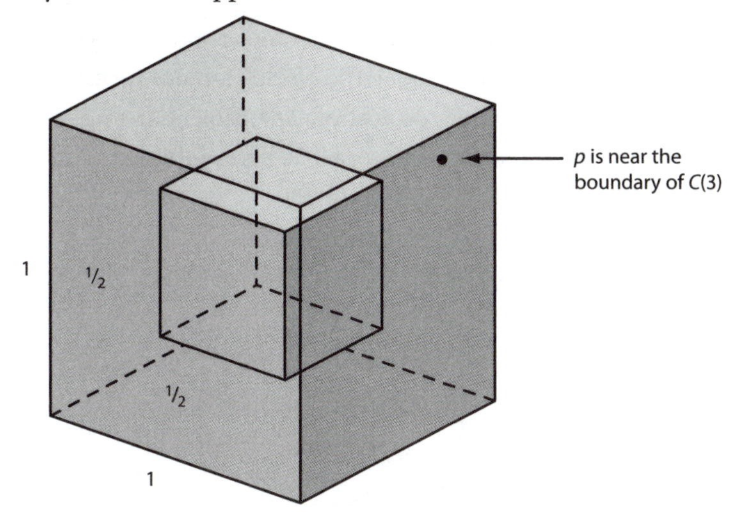

Let $\mathcal{B}(n, r)$ denote the n-dimensional ball of radius r, that is, $\mathcal{B}(n, r)$ is the set of all points in $\mathbb{R}^n$ whose distance from a fixed point—the center—is less than or equal to r. Using analysis, it can be shown that the volume of the ball is given by

$$V(\mathcal{B}(n, r)) = \frac{\pi^{n/2} r^n}{\Gamma(\frac{n}{2}+1)}$$

where $\Gamma(x)$ denotes the *Gamma function*: $\Gamma(x) = \int_0^\infty e^{-t} t^{x-1}\, dt$. We mention (without proof) that for any positive integer n, $\Gamma(n) = (n-1)!$. Thus the Gamma function is, in some sense, a generalization of the factorial function. In this case, we note that for *even* dimensions, say $2m$, the volume formula for a ball becomes

$$V(\mathcal{B}(2m, r)) = \frac{\pi^m r^{2m}}{m!}.$$

13.15. Show that $\mathcal{B}(2m, 1/2)$ can be inscribed in $\mathcal{C}(2m)$. Now compare the volumes of $\mathcal{C}(2m)$ and $\mathcal{B}(2m, 1/2)$ as $m \to \infty$. What do you conclude about the "center" of higher dimensional cubes?

Stepping back

Find a "calculus connection" between the circumference of a circle of radius r and its area. In other words, find a calculus "move" that will generate the area of a circle of radius r simply by using the circumference formula. Give a heuristic argument that shows this connection is sensible. Can you now generalize this notion to volumes and surface areas of spheres in higher dimensions?

14

A synergy between geometry and numbers

Circles and Pythagorean triples

Here we open with a classic result that is arguably one of the most important theorems in mathematics—the Pythagorean Theorem. After we consider the many sides (and the one hypotenuse) of this mathematical gem, we will be inspired to consider the unit circle and to search for rational points on that hypnotically symmetric curve.

Rightful triangles

We now consider a wonderful geometrical means of proving an age-old classic that is attributed to the twelfth-century Indian mathematician Bhaskara.

14.1. Cut out four copies of a right triangle and a square with length equal to the difference between the lengths of the legs of the triangle.

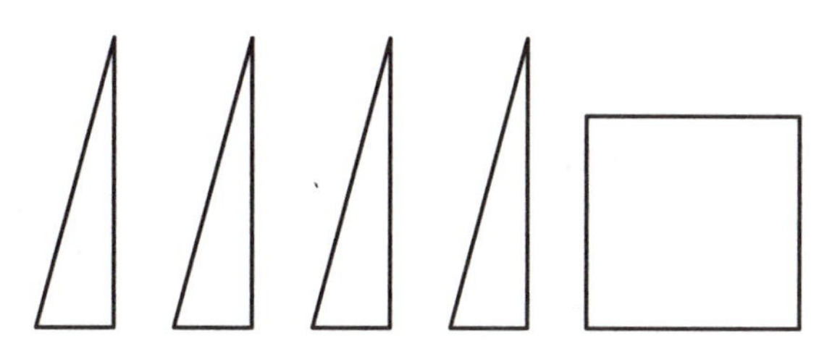

Using these five puzzle pieces, create a square that has sides equal to the triangle's hypotenuse.

14.2. Using the same five puzzle pieces from Challenge 14.1, create *two* squares—one with side length equal to one leg of the original triangle and the other with side length equal to the other leg.

14.3. Using the previous two "puzzling" challenges as inspiration, produce a geometric proof of the following golden oldie:

THEOREM (Pythagorean Theorem). *Given a right triangle with leg lengths a and b and hypotenuse c, then* $a^2 + b^2 = c^2$.

Determining which triangles are allright

Here we explore the converse to the Pythagorean Theorem.

14.4. Prove the following generalization of the Pythagorean Theorem:

THEOREM. *Given the triangle below with sides a, b, and c and altitude h, then*

$$a^2 + b^2 = c^2 + 2aa'.$$

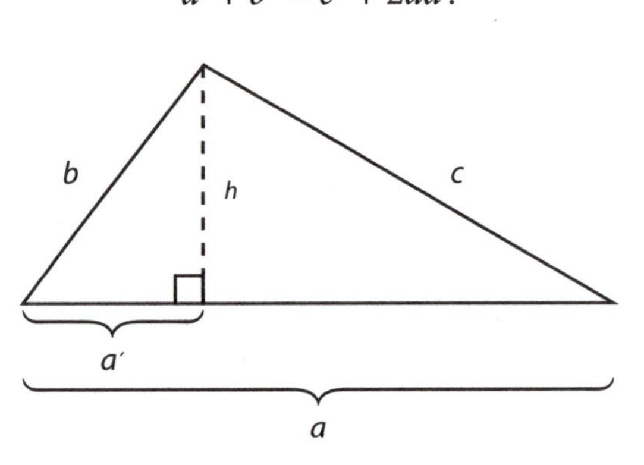

14.5. Prove and extend *or* disprove and salvage:

STATEMENT. *If the sides of a triangle have lengths a, b, and c and if* θ *is the angle between the sides having lengths a and b, then*

$$c^2 = a^2 + b^2 + 2ab\cos\theta.$$

14.6. Is the converse of the Pythagorean Theorem true? That is, can the Pythagorean Theorem be extended to an "if and only if" statement? If so, provide a proof; if not, offer a counterexample.

A rational look at the circle

Inspired by our previous results, we now wonder how to find all *integer* solutions to the equation $a^2 + b^2 = c^2$. The key will be to connect this equation to a simple geometric observation in the xy-plane. This connection is the foundation for a very important modern area known as *arithmetic geometry.* We will further explore this connection in Module 19. In fact this area is home to some of the central ideas that finally led to the proof of Fermat's Last Theorem.

An integer solution (x, y, z) to the equation

$$x^2 + y^2 = z^2$$

is called a *Pythagorean triple.* For example, (3, 4, 5) is a Pythagorean triple. A Pythagorean triple (A, B, C) is called *primitive* if A, B, and C are relatively prime positive integers. We will now apply some ideas of arithmetic geometry to find *all* primitive Pythagorean triples.

We say that a point (x, y) in $\mathbb{R}^2$ is a *rational point* if both coordinates x and y are rational numbers.

14.7. Prove and extend *or* disprove and salvage:

Statement. *Let* $f(x, y, z) = (\frac{x}{z}, \frac{y}{z})$. *Then* f *is a one-to-one, onto map from Pythagorean triples* (x, y, z)*, with* $z \neq 0$*, to rational points on the unit circle*

$$X^2 + Y^2 = 1. \tag{14.1}$$

Through Challenge 14.7, we discover that finding *integer* solutions to $x^2 + y^2 = z^2$, with $z \neq 0$, is equivalent to finding *rational* solutions to $X^2 + Y^2 = 1$. Thus to find all integer solutions to the Pythagorean equation, it is enough to find all rational points on the unit circle. We begin by noting that $(-1, 0)$ is one such solution to (14.1).

14.8. Prove and extend *or* disprove and salvage:

STATEMENT. *If* (x, y) *is a rational solution to* (14.1) *other than* $(-1, 0)$, *then the line through* $(-1, 0)$ *and* (x, y) *has rational slope. Conversely, for any rational number* m, *the line passing through* $(-1, 0)$ *with slope* m *intersects the unit circle at a second point having rational coordinates.*

Thus we are finding a correspondence between the rational numbers (the slopes of lines, written in lowest terms) and rational points on the unit circle. This naive link between arithmetic (rational points) and geometry (slopes of lines) foreshadows the essence of arithmetic geometry.

14.9. Prove and extend *or* disprove and salvage:

STATEMENT. *Suppose that* $m = s/r$. *Then the line of slope* m *passing through* $(-1, 0)$ *also intersects the unit circle at*

$$\left(\frac{r^2 - s^2}{r^2 + s^2}, \frac{rs}{r^2 + s^2}\right).$$

If we multiply through by the denominator, we can describe all primitive Pythagorean triples.

14.10. Prove the following theorem:

THEOREM. *There are infinitely many primitive Pythagorean triples* (x, y, z). *Moreover they are all given by*

$$x = r^2 - s^2, \quad y = 2rs, \quad z = r^2 + s^2,$$

where r *and* s *are relatively prime positive integers such that* $r > s > 0$ *and either* r *or* s *is even.*

Stepping back

Using the methods developed in this module, find all integer solutions (x, y, z) to the equation

$$x^2 + py^2 = z^2,$$

where p is a fixed prime number.

15

The mathematical mysteries within a sheet of paper

Unfolding pattern and structure

Using our mathematical mind-set, we will discover the surprisingly rich structure hidden in the simple process of repeatedly folding a piece of paper.

Getting into the fold

Given a rectangular sheet of paper, we define the *right-folding sequence* to be the result of taking the right edge and folding it on top of the left edge and then repeating this procedure with the same folded piece of paper. By *stage n of the right-folding sequence,* we mean that this paper-folding process has been repeated n times.

Stage 1 of the right-folding sequence

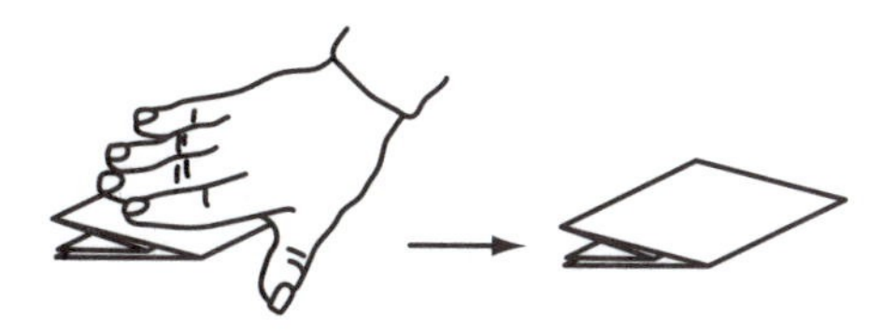

Stage 2 of the right-folding sequence

The folded paper has two components: the *folds* (the creases) and the *strips* (the rectangular segments separated by the folds). We call a fold *right,* denoted R, if the fold resembles $>$, and we say a fold is *left,* denoted L, if the fold resembles $<$.

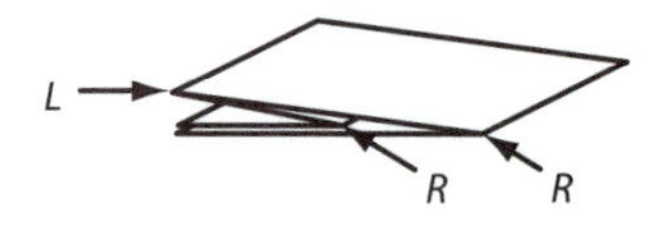

Stage 2 of the right-folding sequence

In the challenges that are about to unfold, you are encouraged to experiment by folding actual paper to develop new insights and build intuition into the arguments you are about to create.

15.1. Suppose we repeated the right-folding sequence a number of times. Show that the fold sequence, starting from the bottom of the folded paper and following the edge of the paper, is of the form R, L, R, L, R, L, ...

15.2. Prove that the number of strips at the nth stage of the right-folding sequence equals 2^n.

15.3. Prove and extend *or* disprove and salvage:

STATEMENT. *The total number of folds at the nth stage of the right-folding sequence equals*

$$2^{n-1} + 2^{n-2} + \cdots + 2^1 + 2^0.$$

15.4. Prove that the number of right folds at the nth stage of the right-folding sequence equals 2^{n-1} and that the number of left folds at the nth stage equals $2^{n-1} - 1$.

The story unfolds

Suppose we have placed a number of right folds into a piece of paper. We now carefully *unfold* the paper so that the bottom strip returns to the left-most side of the paper. The folds—which formerly came in two flavors, R and L—now have two basic shapes: Each is either of the form V, which we call a *valley,* or of the form ∧, referred to as a *ridge*. The sequence of valley and ridge folds, reading from left to right, resulting from n repeated right folds is called the *nth stage of the Life sequence.*

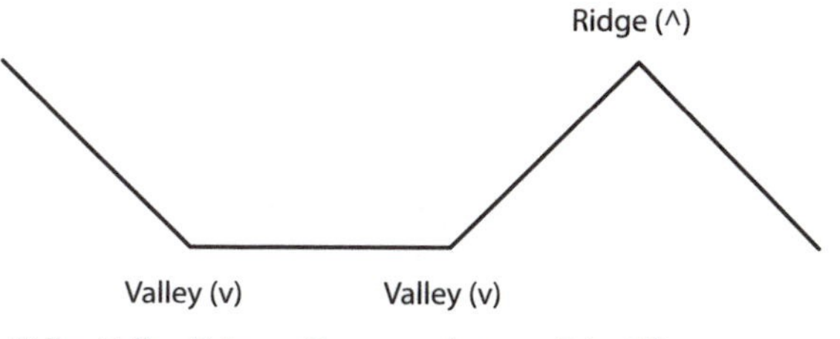

Valley Valley Ridge = The second stage of the Life sequence

(*Bonus puzzler:* Why do we call it the Life sequence? *Answer:* It has its ups and downs.)

15.5. Suppose a piece of paper contains some number of right folds, and we color the *new* folds introduced at the next stage red. Prove that after unfolding the paper, we see that the red folds form the alternating sequence (reading from left to right) ∨, ∧, ∨, ∧, ∨, ∧, ...

15.6. Prove that the nth stage of the Life sequence equals the $(n-1)$th stage of the Life sequence after it has been perfectly shuffled (interwoven) with the alternating sequence ∨, ∧, ∨, ∧, ∨, ∧, ..., where the number of terms in the alternating sequence is 2^{n-1}.

15.7. How do the first $2^{n-1}-1$ folds of the nth stage of the Life sequence compare with the $(n-1)$th stage of the Life sequence? Prove your answer.

15.8. How do the last $2^{n-1}-1$ folds of the nth stage of the Life sequence compare with the $(n-1)$th stage of the Life sequence? Prove your answer.

Adding arithmetic to generate a wild parity

Suppose we now convert the valleys, ∨, to 1's and the ridges, ∧, to 0's. We call the sequence of 0's and 1's that arises from the converted nth stage of the Life sequence the *nth stage of the paper-folding sequence,* denoted PFS_n, and by $PFS_n(m)$, we mean the mth term in the sequence PFS_n.

15.9. For any positive integer m, prove that the $\lim_{n\to\infty} \text{PFS}_n(m)$ exists. If we now write f_m for this limit, why can we view f_m as the mth term in the *infinite* paper-folding sequence?

15.10. Consider the following theorem:

THEOREM. *If the function $\mathcal{F}(x)$ is defined to be the formal power series given by $\mathcal{F}(x) = \sum_{m=1}^{\infty} f_m x^m$, where f_m denotes the mth term in the infinite paper-folding sequence, then*

$$\mathcal{F}(x) - \mathcal{F}(x^2) = \frac{x}{1-x^4}.$$

By writing out the beginning of the formal power series $\mathcal{F}(x)$ and $\mathcal{F}(x^2)$, consider the difference and find a pattern. Assuming your pattern continues, prove the previous theorem.

Stepping back

A simple Turing machine is a machine that can read and write a ticker tape of numbers. The machine starts at the beginning of the list of numbers and reads the first few numbers. Given what it reads, the machine then writes additional numbers at the end of the ticker tape list. The set of rules that instructs the machine what to write given what it has just read is known as a *program*. Can we use a Turing machine to generate the sequence PFS_n for each n? The answer is yes! Suppose we will only read and write digits from 1 to 4. We start with 1 on the ticker tape. The machine will read one digit at a time, and given what it reads, it will write two digits at the end of the list. Create a program (that is, the set of rules for what to write given what is read) so that when the list of digits on the ticker tape is transformed to just 0's and 1's (we convert all even digits to 0's and all odd digits to 1's), the resulting list is precisely PFS_n for each n.

16

Take it to the limit

An initial approach to analysis

At the very heart of calculus is the notion of limit. This central idea involves studying functions, in some sense under a microscope, to establish results at an ultrasubatomic mathematical scale. These themes come together in an area of mathematics known as analysis. Here we introduce some of these basic notions and take a close-up look at the foundations upon which calculus is built.

Getting really close

Let a_1, a_2, a_3, ... be an infinite sequence of real numbers. We say the *limit as n approaches infinity of the sequence* $\{a_n\}$ equals L (or $\{a_n\}$ *converges to* L) denoted $\lim_{n\to\infty} a_n = L$, if for any given $\varepsilon > 0$, there exists an integer N such that for all indices $n \geq N$, it follows that

$$|a_n - L| < \varepsilon.$$

In other words, the terms of the sequence $\{a_n\}$ get arbitrarily close to L as the index n goes off to infinity. We say that the *limit of a sequence* $\{a_n\}$ *exists* if there exists an L such that $\lim_{n\to\infty} a_n = L$. If no such limit L exists, then we say that the *limit of* $\{a_n\}$ *does not exist* or that the sequence $\{a_n\}$ *diverges.*

If $\lim_{n\to\infty} a_n = L$ and we are given an $\varepsilon > 0$, then we can view the parameter N as the "waiting time" required until the elements of the tail end of the sequence $\{a_n\}$, for all $n \geq N$, will be within ε of L. Thus if we select a smaller ε, then we might be required to wait longer, and the associated N might be required to be larger.

To establish that $\lim_{n\to\infty} a_n = L$, we must consider a "generic" positive number, and for that unspecified value, we must produce an N—usually described as a function of ε—that satisfies the definition of limit.

16.1. Let $a_n = \frac{2n+1}{3n-500}$. Does $\lim_{n\to\infty} a_n$ exist? If so, find the limit. Justify your answer using the definition of limit.

16.2. Let $a_n = \frac{n^2-1}{n+1}$. Does $\lim_{n\to\infty} a_n$ exist? If so, find the limit. Justify your answer using the definition of limit.

16.3. Prove and extend *or* disprove and salvage:

STATEMENT. *Suppose that $\{a_n\}$ and $\{b_n\}$ are two infinite sequences. If $\{a_n\}$ converges to A and $\{b_n\}$ converges to B, then $\{a_n + b_n\}$ is a convergent sequence having limit $A + B$. If both $\{a_n\}$ and $\{b_n\}$ diverge, then $\{a_n + b_n\}$ is a divergent sequence.*

16.4. Prove and extend *or* disprove and salvage:

STATEMENT. *Suppose that $\{a_n\}$ and $\{b_n\}$ are two infinite sequences such that for each index n,*

$$a_n \le b_n \le a_n + \frac{1}{n}.$$

If $\lim_{n\to\infty} b_n = B$, then $\lim_{n\to\infty} a_n = B$.

16.5. Prove and extend *or* disprove and salvage:

STATEMENT. *Suppose that $\lim_{n\to\infty} a_n = A$. Let $\{b_n\}$ be the infinite sequence defined by*

$$b_n = \begin{cases} a_n & \text{if } n \text{ is even;} \\ 0 & \text{if } n \text{ is odd.} \end{cases}$$

Then $\lim_{n\to\infty} b_n = A$.

Let $\{a_n\}$ be an infinite sequence of real numbers. We say that $\{a_n\}$ is a *Cauchy sequence* if for any given $\varepsilon > 0$, there exists an integer N such that for all pairs of indices m, n satisfying $n \geq m \geq N$, it follows that

$$|a_n - a_m| < \varepsilon.$$

Thus a sequence is a Cauchy sequence if for any given $\varepsilon > 0$, there is a "waiting time" N such that for all $n \geq N$, the a_n's are within ε of each other. That is, the terms of the sequence are getting arbitrarily close to each other as their indices journey farther and farther out.

16.6. Prove and extend *or* disprove and salvage:

STATEMENT. *If $\{a_n\}$ converges, then $\{a_n\}$ is a Cauchy sequence.*

16.7. Prove and extend *or* disprove and salvage:

STATEMENT. *Let $\{a_n\}$ and $\{b_n\}$ be two infinite sequences satisfying $a_n \leq b_n$ for all $n = 1, 2, 3, \ldots$, and write $I_n = [a_n, b_n]$ for the closed interval from a_n to b_n. If for all n, $I_{n+1} \subseteq I_n$ and* $\text{length}(I_{n+1}) = \frac{1}{2}\text{length}(I_n)$, *then $\{a_n\}$ and $\{b_n\}$ are both Cauchy sequences.*

Grasping the cloudlike Cantor set

We now construct an interesting subset of the interval $[0, 1]$ via an infinite process that possesses a number of counterintuitive properties. We begin with the interval $[0, 1]$ and eventually prune away infinitely many subintervals in a systematic manner. At the first stage of our process, we remove the middle third of the interval, thus leaving us with two disjoint intervals, each of length $\frac{1}{3}$: the first third and the third third of $[0, 1]$.

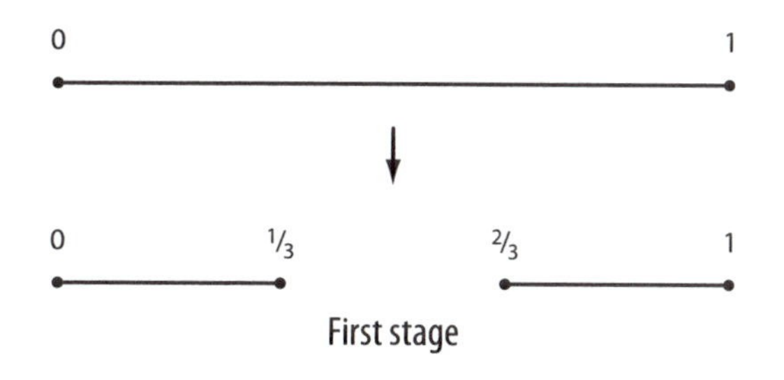

First stage

At the second stage, we remove the middle third of each of the two subintervals, thus leaving us with four disjoint intervals, each of length $\frac{1}{9}$.

Second stage

We can now inductively describe the remaining stages of the construction. We assume that the nth stage of the construction is completed. To generate the $(n + 1)$th stage, we consider each interval from the nth stage and remove its middle-third subinterval.

Third stage

We iterate this pruning process for all $n = 1, 2, 3, \dots$. The collection of points from $[0, 1]$ that are never removed is called the *Cantor set* or, more precisely, the *middle-third Cantor set,* denoted $\mathcal{C}$.

16.8. How many subintervals from the $(n - 1)$th stage are removed to produce the nth stage of the Cantor set construction?

16.9. What is the total length of *all* the intervals *removed* up to and through the nth stage of the Cantor set construction?

16.10. Prove and extend *or* disprove and salvage:

STATEMENT. *The sum of the lengths of all intervals that make up the set* $[0, 1] \setminus \mathcal{C}$ *equals* 1.

Turning to ternary

We traditionally view the real numbers as infinitely long decimal expansions in base 10—thus the allowable digits are 0, 1, 2, ... , 8, 9. Specifically, if α is a point on the real number line, then for some positive integer N, we have $\alpha = \pm\sum_{n=-N}^{\infty} a_n / 10^n$, where the coefficients $a_n \in \{0, 1, 2, \dots, 8, 9\}$.

16.11. Prove and extend *or* disprove and salvage:

STATEMENT. *Given a real number α, there exists a positive integer N such that*

$$\alpha = \sum_{n=-N}^{\infty} \frac{a_n}{3^n}, \text{ where each } a_n \in \{0, 1, 2\}.$$

The expansion given in Challenge 16.11 is known as the *ternary expansion* for α. By viewing the real numbers in base 3, we can discover some beautiful structure possessed by the Cantor set.

16.12. Prove and extend *or* disprove and salvage:

STATEMENT. *Every real number in* $[0, 1]$ *can be expressed uniquely as the sum of two numbers whose ternary expansions are of the form*

$$\sum_{n=0}^{\infty} \frac{a_n}{3^n}, \text{ where each } a_n \in \{0, 1\}.$$

16.13. Prove and extend *or* disprove and salvage (with another "if and only if" statement):

STATEMENT. *A point c is in C if and only if $c = \sum_{n=0}^{\infty} a_n / 3^n$, for which each $a_n \in \{0, 2\}$.*

16.14. Prove and extend *or* disprove and salvage:

STATEMENT. *The Cantor set is a countable set.*

16.15. Prove and extend *or* disprove and salvage:

STATEMENT. *Every real number in the interval* $[0, 2]$ *can be expressed as the sum of two elements of the Cantor set C.*

Stepping back

In view of Challenge 16.10, we discover that the Cantor set is a set having "length" 0. Can you slightly modify the construction of the Cantor set so that after infinitely many steps, the resulting object will have the same structure as the Cantor set but with positive "length"?

17

Uninterrupted thoughts of continuity

A jump-free journey

Continuous functions are a natural starting point for any venture into calculus, because essentially all physical motion is continuous. In our youth, we focused on the visual essence of a continuous function of one variable: $y = f(x)$ is continuous if its graph can be drawn without lifting a pencil (or pen, if we were unleaded). Here we return to the idea of continuity and provide a rigorous and precise mathematical definition. After exploring some important and attractive consequences, we consider a devilish function that exhibits some sneaky behavior.

Functions without lifting a finger

Let $f: \mathbb{R} \to \mathbb{R}$ be a function. We say that f is *continuous at* x_0 if for any given $\varepsilon > 0$, there exists a $\delta > 0$ such that for every $x \in \mathbb{R}$ satisfying $|x - x_0| < \delta$, it follows that $|f(x) - f(x_0)| < \varepsilon$. That is, as x gets close to x_0, $f(x)$ gets close to $f(x_0)$. We say that f is *continuous* if f is continuous at x_0 for *all* x_0 in the domain of f.

17.1. Prove and extend *or* disprove and salvage:

STATEMENT. *For any fixed real numbers m and b, the function $\ell(x) = mx + b$ is continuous.*

17.2. Prove and extend *or* disprove and salvage:

STATEMENT. *If there exists a real number τ such that the function $\tau h(x)$ is continuous, then $h(x)$ is a continuous function.*

17.3. Prove and extend *or* disprove and salvage:

STATEMENT. *If $f(x)$ and $g(x)$ are both continuous functions, then $f(x) + g(x)$ is a continuous function.*

If a function is not continuous at a point x_0, then we say it is *discontinuous* at x_0.

17.4. Prove and extend *or* disprove and salvage:

STATEMENT. *If either $f(x)$ or $g(x)$ is discontinuous at x_0, then $f(x) + g(x)$ is discontinuous at x_0.*

17.5. Prove and extend *or* disprove and salvage:

STATEMENT. *If $f(x)$ is a continuous function and $\{x_n\}$ is a Cauchy sequence with, say, $\lim_{n\to\infty} x_n = L$, then $\lim_{n\to\infty} f(x_n) = f(L)$.*

Valuing intermediate values

We are now ready to prove an extremely important result that has an immense number of consequences.

17.6. Prove the following:

THEOREM (The Intermediate Value Theorem). *Suppose that $f(x)$ is a continuous function on the interval $[a, b]$, with $f(a) < f(b)$. Then for any y satisfying $f(a) < y < f(b)$, there exists an $x \in [a, b]$ satisfying $f(x) = y$.*

17.7. Prove and extend *or* disprove and salvage:

STATEMENT. *Suppose that $\mathcal{K}$ is a circle in the plane and $F : \mathcal{K} \to \mathbb{R}$ is a continuous function. Then there exist two antipodal points $\mathbf{x}_1$ and $\mathbf{x}_2$ on $\mathcal{K}$ such that $F(\mathbf{x}_1) = F(\mathbf{x}_2)$.*

17.8. Must there always exist two antipodal points on Earth having the exact same temperatures? Justify your answer.

17.9. Prove and extend *or* disprove and salvage:

STATEMENT. *Let $P(x)$ be a polynomial with real coefficients. Then there exists a real number x_0 such that $P(x_0) = 0$.*

A devilish staircase

Here we define a function $D: [0, 1] \to [0, 1]$. Our description of $D(x)$ will be analytical and complicated. But once we evaluate D at a few points, we will start to get a sense as to what is really going on. The key is to first recall the Cantor set and the statement from Challenge 16.13.

Given a real number $x \in [0, 1]$, we recall from the previous module that x can be expressed in its ternary expansion as the infinite series

$$x = \sum_{n=1}^{\infty} \frac{a_n}{3^n}, \quad a_n \in \{0, 1, 2\}.$$

Of course, real numbers can be expressed in any base. For example, we could also express $x \in [0, 1]$ in its binary (base 2) expansion as the infinite series

$$x = \sum_{n=1}^{\infty} \frac{b_n}{2^n}, \quad b_n \in \{0, 1\}.$$

To define our function $D(x)$, we first express x in its ternary expansion, $x = \sum_{n=1}^{\infty} a_n/3^n$, where $a_n \in \{0, 1, 2\}$. If none of the a_n's equal 1, then they are all even, and so if we write $b_n = a_n/2$, then we have that $b_n \in \{0, 1\}$ for all n. We now define $D(x) = \sum_{n=1}^{\infty} b_n/2^n$; that is, we define the value $D(x)$ as a *binary* expansion.

If *at least* one a_n equals 1, then we truncate the infinite series *after* the first term for which $a_n = 1$; that is, we consider the sum $\sum_{n=1}^{N} a_n/3^n$, where $a_N = 1$ and $a_n \in \{0, 2\}$ for all $n = 1, 2, \ldots, N-1$. As before, we write $b_n = a_n/2$ for all $n = 1, 2, \ldots, N-1$ and define $b_N = 1$. Again we note that each of these b_n's is equal to either 0 or 1. We now define $D(x)$ in this case by $D(x) = \sum_{n=1}^{N} b_n/2^n$. There is no doubt—this is one complicated function. Now let's develop some insight into this strange beast.

17.10. Compute the following:

- $D(0)$
- $D(1)$
- $D(x)$ for any $x \in [\frac{1}{3}, \frac{2}{3}]$
- $D(x)$ for any $x \in [\frac{1}{9}, \frac{2}{9}]$
- $D(x)$ for any $x \in [\frac{7}{9}, \frac{8}{9}]$

17.11. Give a rough sketch of the graph of $y = D(x)$.
Remark: This graph is known as the *devil's staircase*. Can you guess why?

17.12. Verify that $D(x)$ is an increasing function; that is, if $x_1 < x_2$, then $D(x_1) \leq D(x_2)$.

Suppose that $f: \mathbb{R} \to \mathbb{R}$ is a function. We say that f is *locally constant at* x_0 if there exists an $\varepsilon > 0$ such that for all x satisfying $|x - x_0| < \varepsilon$, it follows that $f(x) = f(x_0)$. That is, f is locally constant at x_0 if there is an open interval surrounding x_0 for which the function f is constant.

17.13. Describe the set of x-values in $[0, 1]$ for which $D(x)$ is locally constant.

17.14. Compute the sum of the lengths of all the intervals that make up the set $\{x \in [0, 1] : D'(x) = 0\}$. (Here $D'(x)$ denotes the usual derivative from calculus.)

17.15. Prove and extend *or* disprove and salvage:

STATEMENT. *If $f: [0, 1] \to \mathbb{R}$ is an increasing continuous function such that $f(0) = 0, f(1) = 1$, then there must exist a nontrivial interval I in $[0, 1]$ for which $f'(x) > 0$ for all $x \in I$.*

Stepping back

Let $f_1, f_2, \ldots$ be a sequence of functions, each mapping the interval [0, 1] into [0, 1]. We say that the *limit of the sequence of functions exists* if there exists a function $f: [0, 1] \to [0, 1]$ such that for each $x_0 \in [0, 1]$, we have that

$$\lim_{n \to \infty} f_n(x_0) = f(x_0).$$

Sometimes this notion is called *pointwise convergence of functions.* Suppose that the functions $f_n(x)$ converge pointwise to the function $f(x)$. If each function $f_n(x)$ is continuous, then must the limit function $f(x)$ be continuous? If so, can you give an explanation? If not, can you offer a counterexample?

18

An abstract world of algebra

Reconciling with your x

In our early youth, we learned how to solve such algebraic equations as $x + 3 = 5$. While at one point in our lives solving such an expression was a monolithic challenge, today we just glance at it and instantly respond $x = 2$. Below, however, we solve that equation in slow motion—showing every detail and highlighting all the algebraic steps required. There are more steps than perhaps meet the untrained mathematical eye.

$x + 3 = 5$	(the beast to be solved)
$(x + 3) + -3 = 5 + -3$	(adding –3, the additive inverse of 3, to both sides)
$x + (3 + -3) = 5 + -3$	(the associative property of addition)
$x + 0 = 5 + -3$	(defining trait of an additive inverse)
$x = 5 + -3$	(defining trait of the additive identity 0)
$x = 2$	(computation using software)

Thus, to solve the simple equations of our childhood, we require an additive identity element, additive inverses, and the associative property for addition. Here we abstract the essence of these arithmetical observations and discover powerful and rich algebraic structure all around us.

Grasping at groups

We begin by abstracting the notion of addition. Let S be a set. We define a *binary operation* $*$ on the set S to be a function $*: S \times S \to S$.

18.1. (a) For each map $*$ below, determine if it is a binary operation on the implied set. If $*$ is not a binary operation, try to modify the set so the map will be a binary operation on the new set. (Note that $\mathbb{Z}^+$ denotes the set of positive integers and $\mathbb{C}$ denotes the set of complex numbers.)

(i) $*: \mathbb{Z} \times \mathbb{Z} \to \mathbb{Z}$ defined by $a * b = a + b$

(ii) $*: \mathbb{Z}^+ \times \mathbb{Z}^+ \to \mathbb{Z}^+$ defined by $a * b = a - b$

(iii) $*: \mathbb{Q} \times \mathbb{Q} \to \mathbb{Q}$ defined by $a * b = ab$

(iv) $*: \mathbb{Q} \times \mathbb{Q} \to \mathbb{Q}$ defined by $a * b = a/b$

(v) $*: \mathbb{R} \times \mathbb{R} \to \mathbb{R}$ defined by $a * b = \sqrt{a+b}$

(vi) $*: \mathbb{R} \times \mathbb{R} \to \mathbb{R}$ defined by $a * b = \sqrt{a^2 + b^2}$

(vii) $*: \mathbb{C} \times \mathbb{C} \to \mathbb{C}$ defined by $a * b = \pm\sqrt{a+b}$

Let $*$ be a binary operation on a set S. An element $e \in S$ is called an *identity element* if for each element $s \in S$, $s * e = e * s = s$.

18.1. (b) For each binary operation (modified or otherwise) above, is there an identity element?

Let G be a set and $*$ a binary operation on G. We call the pair $\langle G, * \rangle$ a *group* if the following three conditions hold:

- The binary operation $*$ is *associative*; that is, for all $a, b, c \in G$, $(a * b) * c = a * (b * c)$.
- There exists an *identity element* $e \in G$; that is, for all $a \in G$, $a * e = e * a = a$.
- Each element in G has an *inverse* in G; that is, for each element $a \in G$, there exists an element $a' \in G$ satisfying $a * a' = a' * a = e$.

18.2. Prove and extend *or* disprove and salvage:

STATEMENT. *The set of integers together with ordinary addition, $\langle \mathbb{Z}, + \rangle$, is a group.*

18.3. Prove and extend *or* disprove and salvage:

STATEMENT. *Let $\langle G, * \rangle$ be a group. Then there are at most two identity elements.*

18.4. Prove and extend *or* disprove and salvage:

STATEMENT. *Let $\langle G, * \rangle$ be a group. If $a, b \in G$, then the algebraic equation $a * x = b$ is solvable for the unknown x in G.*

A group mentality

Here we discover some famous and important groups.

18.5. Prove and extend *or* disprove and salvage:

STATEMENT. *For a fixed integer $n > 1$, let $\oplus$ denote addition modulo n. Then $\langle \mathbb{Z}, \oplus \rangle$ is a group.*

18.6. Prove and extend *or* disprove and salvage:

STATEMENT. *Let $\langle G_1, *_1 \rangle$ and $\langle G_2, *_2 \rangle$ be two groups. Define $* : (G_1 \times G_2) \times (G_1 \times G_2) \to (G_1 \times G_2)$ by $(a_1, a_2) * (b_1, b_2) = (a_1 *_1 b_1, a_2 *_2 b_2)$. Then $\langle G_1 \times G_2, * \rangle$ is a group.*

Suppose that $\langle G, * \rangle$ is a group. We say that G is an *abelian group* if the binary opearation $*$ is commutative; that is, if for all elements $a, b \in G$, $a * b = b * a$.

18.7. Prove and extend *or* disprove and salvage:

STATEMENT. *Let $\langle G, * \rangle$ be a group. If G is an abelian group, then for all $a, b \in G$, $(a * b)' = b' * a'$.*

18.8. Prove and extend *or* disprove and salvage:

STATEMENT. *Let $\mathbb{Z}_n = \{0, 1, 2, \ldots, n-1\}$. The nonzero elements of $\mathbb{Z}_n$ form a group under multiplication modulo n.*

Building a group table

Suppose that $\langle G, * \rangle$ is a group with $|G|$ (the cardinality of G) finite. In this case, we call G a *finite group.* Then we can produce a table that shows the binary operation $*$ for all pairs of elements of G. Specifically, if $G = \{g_1, g_2, \ldots, g_N\}$, then we can fill in a table in which we place the result of $g_m * g_n$ in the mth row and nth column.

$*$	g_1	g_2	g_3	$\cdots$	g_n	$\cdots$	g_N
g_1							
g_2							
$\vdots$							
g_m					$g_m * g_n$		
$\vdots$							
g_N							

A group table for $\langle G, * \rangle$

18.9. Produce the group tables for $\langle \mathbb{Z}_4, \oplus_4 \rangle$ and $\langle \mathbb{Z}_7, \oplus_7 \rangle$, where the binary operations $\oplus_4$ and $\oplus_7$ denote addition module 4 and 7, respectively. What visual attribute must the group table for a group possess to ensure that the group is an abelian group?

18.10. Is it possible for the same element to appear twice in either a particular column or a particular row of a group table?

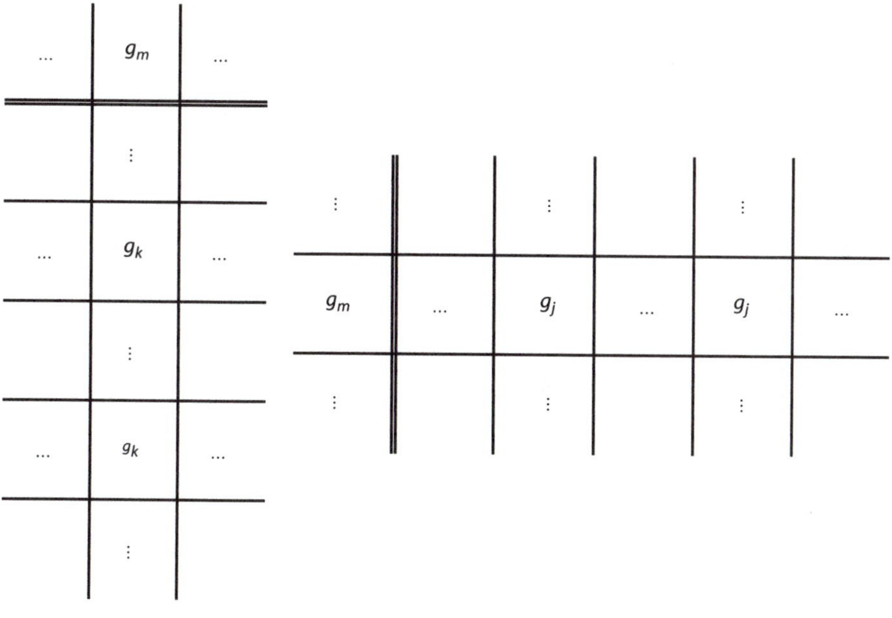

Can either of these happen in a group table for a group $\langle G, * \rangle$?

If so, give an example of such a table and verify that the elements, together with the binary operation, form a group. If not, then prove that such a double feature is impossible.

Stepping back

Using group tables, find all possible distinct groups having four elements. That is, let $G = \{g_1, g_2, g_3, g_4\}$, and let us assume that g_1 is the identity element. Now create binary operations on G that make G a group. Make a group table for each operation and use it to verify that the operation does indeed make G a group. How many different operations can you find?

19

Cycles and curves

Algebraic structure in numbers and geometry

Here we explore some wondrous interconnections among groups, number theory, and geometry. These connections are central and begin to hint at the importance of the notion of a group from abstract algebra.

The subculture of subgroups

Suppose that $\langle G, * \rangle$ is a group. A subset $H \subseteq G$ is called a *subgroup of* G if $\langle H, * \rangle$ is a group.

19.1. Prove and extend *or* disprove and salvage:

STATEMENT. *Let* $\langle G, * \rangle$ *be a group and* H *be a subset of* G. *Then* $\langle H, * \rangle$ *is a subgroup of* G *if and only if* $a * b' \in H$ *for all* $a, b \in H$.

19.2. Prove and extend *or* disprove and salvage:

STATEMENT. *Let* n *be an integer and* $n\mathbb{Z} = \{nt : t \in \mathbb{Z}\}$. *Then* $\langle n\mathbb{Z}, + \rangle$ *is a subgroup of* $\langle \mathbb{Z}, + \rangle$.

19.3. When is $m\mathbb{Z}$ a subgroup of $n\mathbb{Z}$? How does your answer connect group theory to number theory?

Cyclic groups

Let $\langle G, * \rangle$ be a group. For a positive integer n, we let a^n denote the element

$$\underbrace{a * a * \cdots * a}_{n \text{ times}}.$$

We extend this notation as follows: We let a^0 denote the identity element in G and a^{-n} denote $(a')^n$, where a' is the inverse of a in G. We say a group $\langle G, * \rangle$ is a *cyclic group* if there exists an element a such that every element in the group G can be expressed as a^n for some integer n. In that case, the element a is called a *generator* of the group.

19.4. Show that $\mathbb{Z}_n$ is a cyclic group under addition modulo n.

19.5. Prove and extend *or* disprove and salvage:

STATEMENT. *If $\langle G, * \rangle$ is an abelian group, then it is a cyclic group.*

19.6. Is $\mathbb{Z}_4 \times \mathbb{Z}_2$ a cyclic group under pairwise addition (modulo 4 and 2, respectively)? Is $\mathbb{Z}_4 \times \mathbb{Z}_3$ cyclic under its natural binary operation? Make a conjecture of when $\mathbb{Z}_m \times \mathbb{Z}_n$ is a cyclic group. Is there a connection with number theory ideas?

19.7. Prove and extend *or* disprove and salvage:

STATEMENT. *Every subgroup of $\langle \mathbb{Z}, + \rangle$ is a cyclic group.*

A circular group

There is an amazing connection between algebra and geometry. This important interplay—which was briefly foreshadowed in Module 14—leads to an area known as algebraic geometry and to the study of elliptic curves. Here we will give just a hint at this connection by considering the points on a circle and discovering that those points can form a group.

Given a circle $\mathcal{C}$ in the plane, we select and fix a point $\mathcal{O}$ on $\mathcal{C}$.

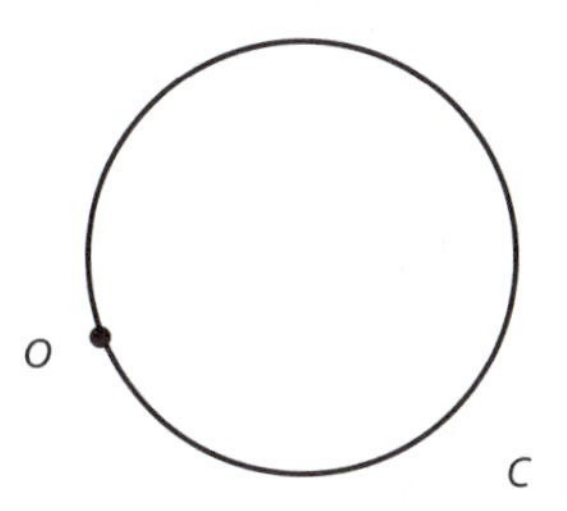

We now define the binary operation $\oplus$ on $\mathcal{C}$ as follows: Given two points P and Q on $\mathcal{C}$, let L_{PQ}

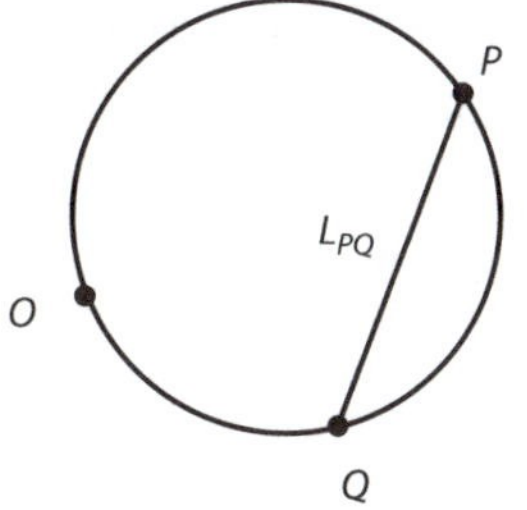

denote the chord between P and Q, and let L denote the line through $\mathcal{O}$ that is parallel to L_{PQ}.

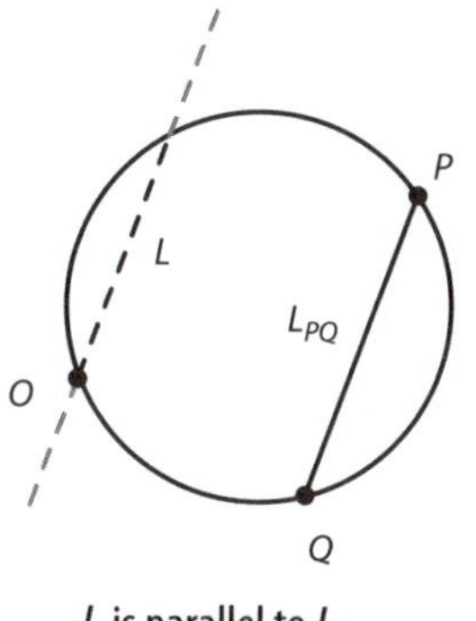

L is parallel to L_{PQ}.

19.8. Show that L will intersect $\mathcal{C}$ at a second point (with the understanding that if L is tangent to $\mathcal{C}$, then the second point of intersection is also $\mathcal{O}$).

We now define $P \oplus Q$ to be this second point of intersection of L and $\mathcal{C}$.

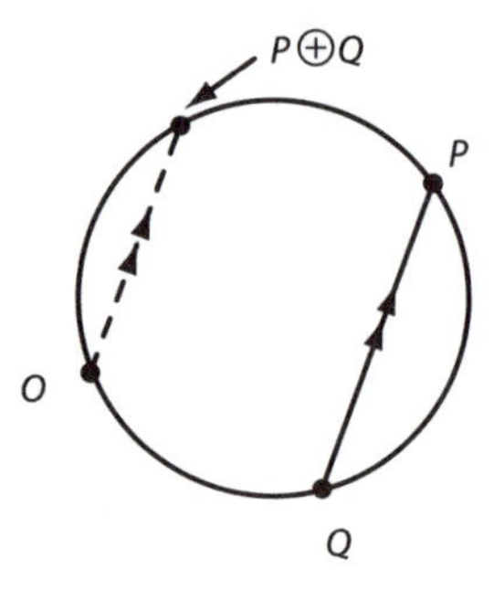

In the case that P and Q are the same point, the chord L_{PQ} is replaced by the line tangent to the circle at P.

Given a point $P \in \mathcal{C}$, we associate with it the unique point $Q \in \mathcal{C}$ such that the line through $\mathcal{O}$ parallel to the chord L_{PQ} is tangent to $\mathcal{C}$. We call this point P^{-1}.

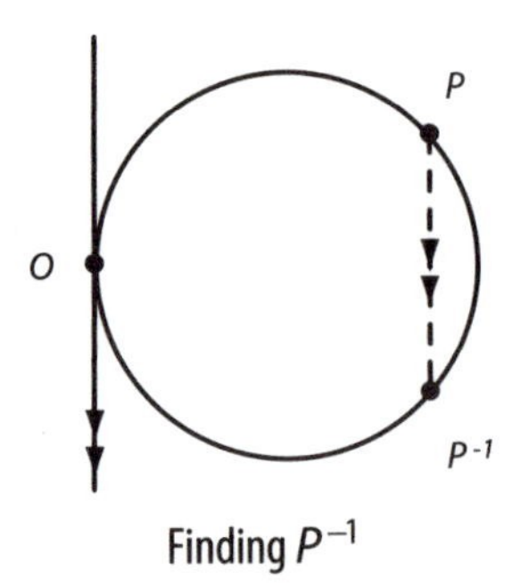

Finding P^{-1}

19.9. Prove that for all points P, $P \oplus \mathcal{O} = \mathcal{O} \oplus P = P$ and $P \oplus P^{-1} = P^{-1} \oplus \mathrm{P} = \mathcal{O}$.

19.10. Prove and extend *or* disprove and salvage:

STATEMENT. *Let G denote the set of points on the circle $\mathcal{C}$. Then $\langle G, \oplus \rangle$ is a group.*

Stepping back

Using the binary operation we found for the set of points on a circle as inspiration, find an analogous binary operation for any conic section in the plane and show that the set of points on the graph of that conic section forms an abelian group.

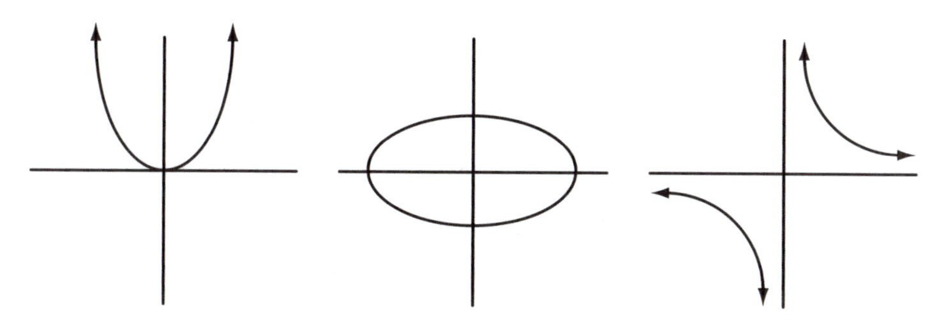

20

Further frontiers

Our journey through this text offered an invitation to readers to be wildly creative and to discover and generate mathematical truths on their own. While essentially all the challenges we have considered have already been resolved within the mathematics community, in reality, most mathematical questions remain open—that is, their answers remain a mystery to all. To celebrate the vast unknown universe of mathematics, we end with ten challenges. Mathematicians are able to answer a number of the following questions, but other questions remain unresolved. Give the following challenges a try. Perhaps you will create a devilishly clever solution to a question that has eluded all mathematicians who have come before you!

Whether on these pages or beyond, at the very center of mathematics is the quest for discovering new and interesting mathematical truths. That quest—as you now well know—is filled with a roller coaster of emotions—from frustration to jubilation. That intellectual journey requires creativity, imagination, and, perhaps above all, tenacity. Good luck and enjoy the exciting challenge that awaits you along the frontiers of mathematics.

20.1. Let **TF** be the set of natural numbers that are "Three-Free"—that is, numbers that do not contain the digit 3. So,

$$\mathbf{TF} = \{1, 2, 4, 5, 6, 7, 8, 9, 10, 11, 12, 14, \dots, 20, 21, 22, 24, \dots, 29, 40, 41, 42, 44, \dots\}\ .$$

Does

$$\sum_{n \in \mathbf{TF}} \frac{1}{n}$$

converge or diverge? If it converges, can you give an explicit upper bound for the sum?

20.2. Can every even number greater than 2 be expressed as the sum of two prime numbers?

20.3. Let $n > 2$ be an integer. Do there exist positive rational numbers r and s such that $\sqrt[n]{r^n + s^n}$ is rational? (*Remark:* Feel free to use any huge theorems whose proofs were front-page news in the mid-1990s.)

20.4. Consider the "formal" (that is, ignore any convergence issues) product

$$\prod_{p \text{ a prime}} \left(1 - \frac{1}{p}\right)^{-1}.$$

If that product were to be expanded out and expressed as an infinite *series*, what would it equal? What does the product imply about the set of prime numbers?

20.5. The middle-thirds Cantor set is a particular Cantor set; there are others. Here is a nonrigorous generic procedure to construct a general Cantor set: We start with a nice set in $\mathbb{R}^N$ (let's think of it as a bounded, connected blob of space).

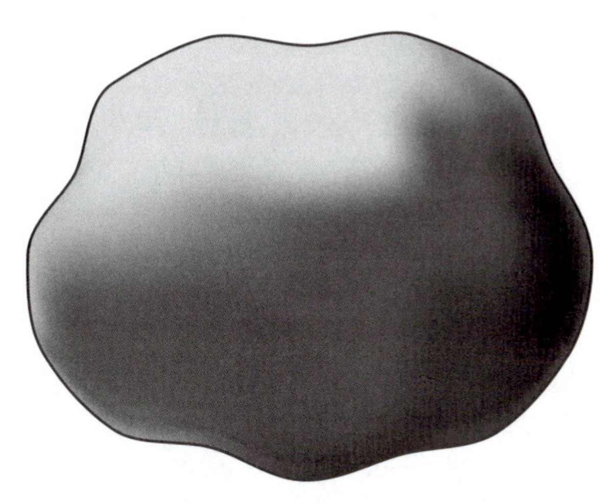

We then select a finite number of disjoint subsets of our original set with the additional property that their "diameters" (length of the blobs) are smaller than the diameter of the original blob.

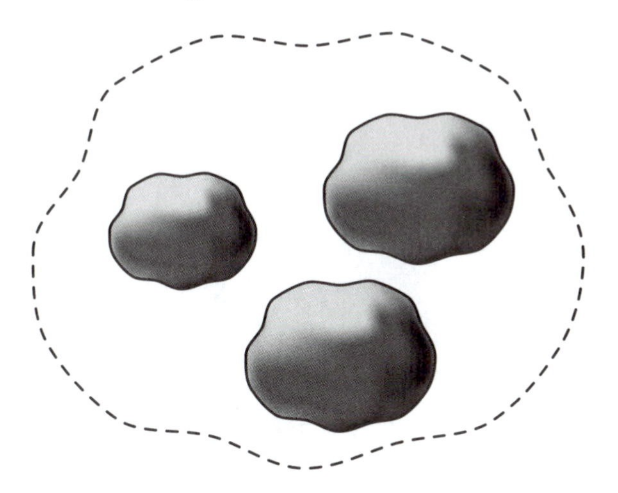

We now repeat: Inside each sub-blob, we select a finite number of sub-sub-blobs with the properties that their diameters are less than that of the sub-blobs.

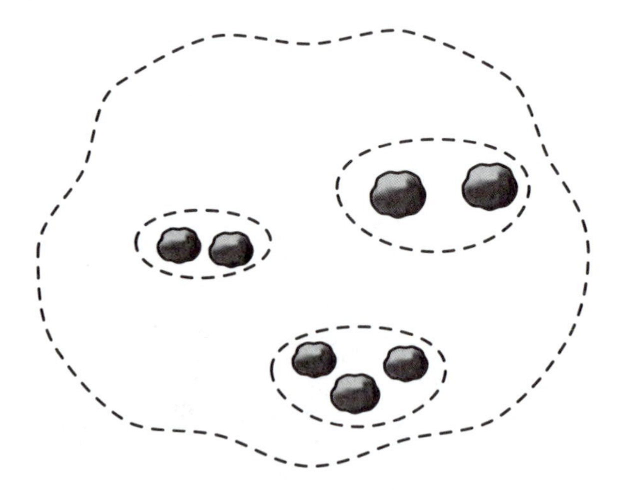

If we continue this procedure for infinitely many steps *and* ensure that the diameters of the blobs are shrinking to 0, then the set we realize in the limit is a general Cantor set. Is it possible to start with a solid unit cube as the initial blob

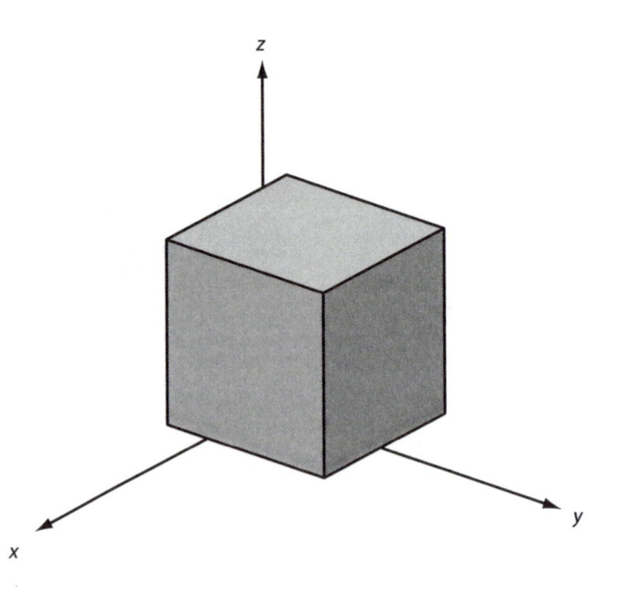

and construct a Cantor set with the property that if a light were to shine down on the Cantor set, the shadow it would cast on the xy-plane would be a solid square?

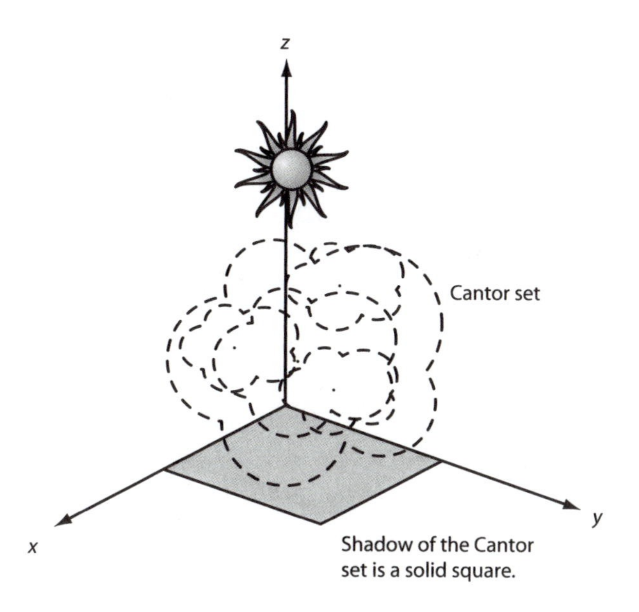

20.6. In the RSA encryption scheme, is breaking the code equivalent to factoring n?

20.7. Consider the following algorithm: Start with a natural number N. If N is even, we divide it by 2; if N is odd, we triple it and then add 1. We take this new number to be our N and repeat the procedure. The process stops if we generate a 1. For example, if we begin with $N = 17$, then we would produce:

$$17, 52, 26, 13, 40, 20, 10, 5, 16, 8, 4, 2, 1\,.$$

Can you prove that for any starting N, this procedure will always end with a 1?

20.8. Let $N > 1$ be a natural number. We say that N is a *perfect number* if N equals the sum of all its proper divisors. For example, 6 is a perfect number because $6 = 1 + 2 + 3$. Find the smallest odd perfect number.

20.9. Are there arbitrarily large gaps between consecutive primes? More specifically, for any given positive integer N, do there exist two *consecutive* prime numbers p, q, $p < q$, such that $q - p \geq N$?

20.10. Let P be a polygonal closed curve with mirrors along its edges. Can you show that there must always exist a point inside P from which every interior point of P is visible?

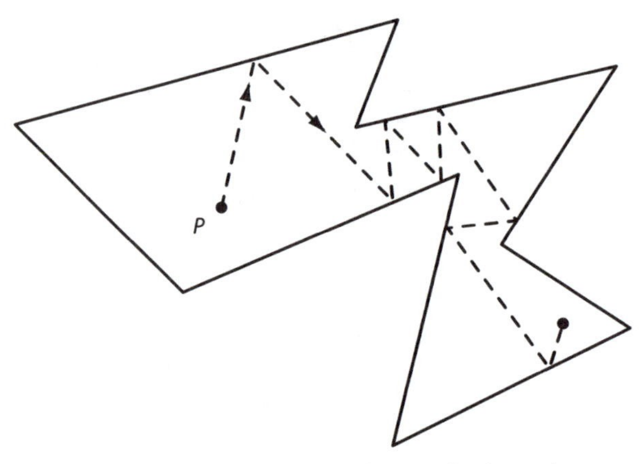

Seeing around corners in a mirrored-covered polygonal room.

A1

Hints, remarks, and leading questions

1.3. Notice that each domino covers two adjacent squares. What can you conclude about the number of covered black squares as compared with the number of covered white squares?

1.4. Try it! How can you fill the endless checkerboard, as described, with natural numbers? Try to extend the example given in the challenge. *Additional remark:* Hidden in this puzzle is an important theorem from analysis involving "harmonic functions" defined on a domain that attain a maximum value along the boundary of the domain. If you are curious, find out what a harmonic function is and hunt down the theorem that is hinted at here. You will then see how it applies to the puzzle at hand.

1.6. First consider the special case in which there are as many heads as there are tails. Even simpler, what if there were only two coins? Generalize from here.

1.7. You can put more than one ball on each side of the balance scale.

1.8. The best way to show there is at least one ball in the barrel is to explicitly state what number is printed on it. Remember, infinity is not a number.

1.9. What if Carol and Chris did not have so many pairs of friends?

1.10. Just remember, the students see all the hats in front of them. The answer is truly surprising.

2.1. *Bonus puzzler:* Read "wow" backward.

2.2. Careful—"The number x is greater than or equal to 0" is not a negation of the first statement. We do not know that x is a real number. Even if we did, we would have to use additional properties of the real numbers to make the deduction that the negation of the original statement implies the statement above. Negating statements that begin "For all . . ." or "There exist . . ." is tricky but extremely important.

2.3. Just to get you going, here is the truth table for Not(A and B).

A	*B*	*A* and *B*	Not (*A* and *B*)
T	T	T	F
T	F	F	T
F	T	F	T
F	F	F	T

2.15. Assume the identity holds for n, and then consider the $n + 1$ case. Let $A_n = a_n + \frac{1}{a_{n+1}}$.

2. Stepping back If there is a formula or method involving the natural numbers for generating the numbers that had the certain property, then induction might be a good proof strategy to consider. Otherwise, it is often best to proceed by contradiction: Assume there are only finitely many numbers possessing that property and use that assumption to deduce a contradiction.

3.2. If d is a common divisor of m and n, then how does d relate to r?

3.3. Can you show that every integer n is between two consecutive multiples of m?

3.5. There are two statements to establish: The process terminates in a finite number of steps, and $\gcd(m, n) = r_{L-1}$.

3.7. Start with the last equality of the Euclidean Algorithm, $r_{L-2} = r_{L-1}q_L$, and use it to rewrite the equality that appears immediately before it so that there is no r_{L-2} term. Apply this idea to the other equalities.

3.9. For this challenge and the (salvage) for Challenge 3.8, can you describe *all* integer solutions? In life, whenever you are given that two numbers are relatively prime, the very first fact you should (must?!) recall is the one found here in 3.9! *Always!*

3.10. If the statement is true, can you weaken the $k \nmid m$ hypothesis? If the statement is false, can you strengthen the $k \nmid m$ hypothesis?

3.12. To prove a result whose conclusion is of the form "Either A or B," assume that one statement (e.g., A) is *false*. Use that assumption to establish the truth of B.

3.15. Suppose there were only finitely many primes. Combine them to produce a number, and then consider Challenges 3.14 and 3.13.

3. Stepping back Consider the collection $\mathcal{E}$ of even positive integers.

4.2. If $a \equiv b \bmod m$, let r_a and r_b be the remainders when a and b are divided by m, respectively. Then show that $-m < r_a - r_b < m$. What are the only multiples of m in that interval? The usefulness of this strategy cannot be overstressed!

4.3. Can you describe all such sets $\mathcal{S}$ for which the result holds?

4.4. Can you extend either conclusion to an "if and only if" statement?

4.6. Convert the congruence into a linear Diophantine equation.

4.7. The statement is known as the Chinese Remainder Theorem.

4.8. Assume two of the numbers are congruent. Now consider the hint for Challenge 4.2.

4.9. When you consider the elements of each set modulo p, how different are the two collections? Consider some specific examples if need be.

4.10. Multiply the elements of each of the two sets from Challenge 4.9. How do those two quantities compare with each other mod p? Now consider your work in Challenge 4.5.

4.12. First show that 3 divides the quantity and then show that 5 divides that value. Finally show that 15 divides the number.

4.14. One strategy to show that $D = W$ is to first show that $D \equiv W \bmod n$. Then using the upper and lower bounds on the numbers D and W, show that it must be the case that $D = W$. To show that $D \equiv W \bmod n$, it is enough to show that $D \equiv W \bmod p$ and $D \equiv W \bmod q$ (Why?). To show these congruences, first verify that

$$D \equiv C^d \equiv W^{ed} \bmod p.$$

Now use the fact that $ed = 1 + my$, together with the definition of m and Fermat's Little Theorem, to establish that $D \equiv W \bmod p$. It might seem as though this hint does all the work, but once you try to work through this outline, you'll see that filling in all the details to turn this overview sketch into an ironclad proof involves considerable creative effort.

4.15. There is one thing the real Alice knows that no one else knows. Can you think of a sneaky way to use that thing to encode a "signature" for the end of her message? What if you switched the roles of the numbers?

5.2. Be careful!

5.3. Assume that $\sqrt{2}$ is rational and consider your work in Challenge 5.2.

5.7. This challenge is truly tricky. Consider two cases with particular values of τ_1 and τ_2: $\tau_1^{\tau_2}$ is irrational, or $\tau_1^{\tau_2}$ is rational. If it's irrational, then take that quantity and raise it to another power.

5.8. Bound the infinite series by a geometric series that converges.

5.10. Assume that e is a rational number with denominator s. Apply Challenge 5.9 to conclude that $s > 1$. Now consider the series $\sum_{n=s+1}^{\infty} 1/n!$. Give upper and lower bounds for this quantity; use those bounds to produce a positive integer less than 1.

6.1. There is a difference between being a subset and being an element of a set.

6.10. The salvage of this result inspires us to extend the definition of "same cardinality" to infinite sets. That is, we say that two infinite sets A and B have the *same cardinality* if and only if there exists a one-to-one, onto function $f: A \to B$.

6.11. Note that elements of a power set are themselves sets—in fact, they are subsets of the original set. Thus there are two elements that are always present in a power set. What are they?

6.12. Recall that $\{0, 1\}^n$ is the set of all ordered n-tuples of 0's and 1's. Also note that one way of describing an element S of a power set $\mathcal{P}(A)$ is to consider each element $a \in A$ and ask: "Is a in S?" The answer is either yes or no.

6.15. Follow the proof for the Schroeder-Bernstein Theorem for the two specific sets. (You may want to consider g to be the identity map and to think of f as mapping x to $\frac{1}{2}x$.)

7.1. Feel free to use either definition of "infinite set." If you use the first, then assume the set $\mathbb{N}$ is finite. Thus there exists a one-to-one, onto function from a finite set to $\mathbb{N}$. Show that such a function cannot exist.

7.2. You can consider your function as a "shift map." Why?

7.3. You can consider your function as a "shuffle map." Why?

7.4. If you can systematically *list* all the rational numbers, then you can find a one-to-one, onto function between the natural numbers and the rationals. You might want to combine the "shift map" (Challenge 7.2) and the "shuffle map" (Challenge 7.3) notions.

7.6. Of course, $|\{1, 2, \ldots, N\}| = N$ and $|\mathcal{B}_N| = 2^N$, so plainly these quantities are not equal—indeed, $|\{1, 2, \ldots, N\}| < |\mathcal{B}_N|$. Unfortunately this simple proof is explicitly not permitted. Thus you need a more elaborate argument—one that will allow you to prove a number of important generalizations. To prove that these sets do not have the same cardinality, you must show that no one-to-one, onto function between these sets can exist. Establish this claim by considering an arbitrary function, $f: \{1, 2, \ldots, N\} \to \mathcal{B}_N$, that is one-to-one. Now construct an element of $\mathcal{B}_N$ that is not in the image of the map to show that the map cannot be onto. This requires a systematic method that will work for an arbitrary function. It might be helpful to consider a specific example first—say, $N = 3$.

7.7. If $\mathcal{B}_\infty$ is countably infinite, there is a one-to-one, onto function between $\mathbb{N}$ and $\mathcal{B}_\infty$. That is, you can list the natural numbers in one column and the elements of $\mathcal{B}_\infty$ in the next column, and every element of $\mathcal{B}_\infty$ should appear in that second column. Can you show that every element of $\mathcal{B}_\infty$ appears on this list or can you find an element of $\mathcal{B}_\infty$ that is missing?

7.12. This question is the hardest challenge from this module. In fact, if you cannot answer it, don't feel too bad—you're not alone. (Ask any mathematician for the surprising explanation!)

7.13. Let $\mathbb{Z}[x]$ denote the set of all polynomials in x with integer coefficients. Show that the set $\mathbb{Z}[x]$ is countable, and recall that each polynomial has a finite number of zeros.

7.14. What is the definition of the number t? How many letters did you use?

8.1. The last two sequences may be a bit easier than the first two.

8.3. For the first one, recall that a is a constant; for the second, notice that

$$\frac{x}{1+x} = x\left(\frac{1}{1+x}\right).$$

For the last one, it might be best to write out the product:

$$\frac{1}{(1-x)^2} = \left(1+x+x^2+x^3+\cdots\right)\left(1+x+x^2+x^3+\cdots\right).$$

8.4. Factor the denominator first. Then consider your results from Challenge 8.3.

8.6. Given $a_n - 2a_{n-1} + a_{n-2}$, you might be inspired to consider

$$f(x) - 2xf(x) + x^2 f(x) = 0.$$

8.9. You will have to apply the quadratic formula to factor the quadratic you produce; that is, the quadratic will have irrational zeros.

9.1. A useful extension in mathematics is one that hypothesizes that there are infinitely many objects to be placed in finitely many boxes. What can you conclude?

9.3. It is trivial to prove the statement as stated. The heart of this challenge is to craft the more interesting extension and then prove it! Your mastery of congruences might be put to good use here.

9.4. If you attempt to salvage (or extend) this statement, consider modifying S, keeping the rest of the statement as it is.

9.13. Left-justify the triangular list of numbers. If you look along certain diagonals, you might discover a famous sequence seen at several points through this text.

9.14. Use induction, together with Challenge 9.12 and a careful change of variables of the indices.

9.15. Think of a good choice for the values a and b, and you're home free!

9. Stepping back The image resembles a well-known fractal.

10.7. First find the probability that there is no match.

10.10. Consider the hint for Challenge 10.7.

11.1. Can you give an explicit formula for the sum?

11.9. What if the graph were made of two disjoint pieces?

11.11. Salvage and then use induction.

11.13. How many such pairs could there be?

12.2. Try induction on the number of circuits in G, keeping in mind that this number could be as small as 0. You'll need results from Module 11. For the extension, consider the converse of the statement. Is it true?

12.4. Check the inequality with $\mathcal{K}_4$ and with $\mathcal{K}_5$ minus one edge. Salvage it with a small change in one number. To get the Euler Characteristic involved in the proof, look at a planar drawing of G, walk around the boundary of each region, and count edges as you go.

12.6. Again count edges on the boundary of each region. Remember G is bipartite.

12.11. Consider $\mathcal{K}_5$ with some extra vertices added to an edge.

12.13. Look for a $\mathcal{K}_{3,3}$.

12.15. Nine is too many. Look at a planar drawing of a regular planar graph G. Express the number of regions in terms of the number of vertices and edges. Let s be the number of edges bounding each region. Show that $s \geq 3$. Now express F in terms of E and s, and express V in terms of E and the vertex degree. Consider all the possible cases that work.

12. Stepping back Yes, the result can be extended. All that is needed is a condition on the relative degrees of the vertices. Can you figure out what that condition is?

13.2. This question can be connected to Statement 12.10 if you just consider the edge-skeleton of the polyhedra and imagine stretching the skeleton so it projects onto the plane. This idea is formally known as *stereographic projection.*

13.10. The Pigeonhole Principle is always a useful idea to consider in various scenarios.

13.14. It helps to think about the volume of an n-dimensional cube. Extending the usual formula for volume, we get the volume of $C(n)$ to be $1^n = 1$ and the volume of an n-dimensional cube with edge length $\frac{1}{2}$ to be $(\frac{1}{2})^n$.

13. Stepping back Consider a continuum of concentric circles, and then compare the circumference with the difference in the areas of two disks having nearly the same radii.

14.2. The two squares might share part of a side.

14.7. Suppose that $x^2 + y^2 = z^2$, with $z \neq 0$. What happens if you divide by z^2? Is there really a one-to-one correspondence? What if x, y, and z were all even?

14.10. The additional arithmetical conditions ensure the solution is primitive.

15. Stepping back As a postscript, search for the phrase "dragon curve" on the Internet and see what you find. Can you see how the fractal curve comes from the paper-folding process?

16.1. Given ε, to find an N that would work (if such an N exists), find the smallest n satisfying

$$\left|\frac{1003}{9n-1500}\right| < \varepsilon.$$

More important, figure out where that inequality came from.

16.2. If you don't think a limit exists and wish to prove that conjecture, then show that for any constant L, you can find an $\varepsilon > 0$ such that no such N satisfying the definition of limit can exist.

16.3. Be careful with the second part of the statement.

16.6. Recall the triangle inequality ($|x+y| \leq |x|+|y|$) and consider

$$|a_n - a_m| = |a_n - L + L - a_m|.$$

16.10. The sum of the lengths of all those intervals is actually a geometric series (waiting to be summed!).

16.11. Find the two consecutive integers that straddle α and write the smaller one in base 3. Now take the interval of length 1 and divide it into three equal subintervals (label them 0, 1, and 2). Now consider the third in which α resides. Take that interval and cut it up into three equal subintervals. Repeat.

16.12. Remember that $2 = 1 + 1$. This trivial remark is the key to this challenge!

16.15. If $\alpha \in [0, 2]$, then $\alpha/2 \in [0, 1]$. Now express $\alpha/2$ in its ternary expansion. Can you write that number as the sum of two numbers whose ternary expansions contain only the digits 0 and 1? What is the next step?

16. Stepping back The generalization of "length" of a set (when the set is not an interval) is called "measure." Your challenge here is to create a set that has positive measure.

17.2. This one is tricky. Can you think of a τ that might cause problems?

17.4. Be careful. Remember that sometimes two wrongs make a right; or in this context, maybe two bad functions can work together to make a good function.

17.6. Let $a_1 = a$, $b_1 = b$, and m denote the midpoint of $[a_1, b_1]$. If $f(m) = y$, then the proof is completed. If $m < y$, then let $a_2 = m$, $b_2 = b_1$, and $I_2 = [a_2, b_2]$. Finally, if $y < f(m)$, let $a_2 = a_1$, $b_2 = m$, and $I_2 = [a_2, b_2]$. Repeat this process and consider the results of Challenges 16.7 and 17.5.

17.7. For a point x on the circle, let $\overline{x}$ denote its antipodal point on the circle. Consider the function $G : \mathcal{K} \to \mathbb{R}$ defined by $G(x) = F(x) - F(\overline{x})$. Show that G is continuous. If $G(x) \neq 0$ for some x, then compare $G(x)$ and $G(\overline{x})$. Now what?

17. Stepping back This is an extremely important question in analysis. The most common guess to this question is incorrect, so you might want to think twice before giving your final answer.

18.1. (a) Fewer than half of those bad boys are actually binary operations! Handle with care.

18.7. Recall that $(a * b)'$ is the inverse in G of the element $a * b$. Also, some food for thought: Is the converse of Statement 18.7 true?

18. Stepping back The notion of distinct groups is an important one in abstract algebra. Of course, if we have a group G and just change the names of the elements but keep the binary operation the same, then even though the group "looks" different, it really is the same group. This idea is formalized in the concept of an *isomorphism*. Roughly speaking, two groups are isomorphic if there is a way to rename the elements so that the group tables of the two groups look identical. In this final challenge of the module, you are asked to generate as many genuinely different groups (ignore a group that is really just a renaming of an already-found group) as possible having four elements.

19.10. The most challenging step is to verify that $\oplus$ is associative. A sneaky way to establish the associativity of $\oplus$ is to find an equivalent description of the binary operation $\oplus$ that depends on angular measurement from the center of the circle to $\mathcal{O}$.

20.4. Given the result found in this challenge, could there be only finitely many primes? Give an alternate proof of the infinitude of the primes.

A2

A proof primer

The proof is in the pudding

Getting the proof out of the pudding and onto the paper

If you have never written formal mathematical proofs, the prospect of such writing is daunting. In a traditional course in which the instructor lectures, students often learn the process of proof by studying the instructor in action. If a course is purely inquiry-based, however, then the instructor barely opens his or her mouth, which while most of the time is awesome, sometimes can be frustrating. Here we offer a few pointers to help you create, craft, and produce a mathematical proof. Remember that all the suggestions that follow are just that—suggestions. As in any writing exercise, it is important for the writer to find his or her own voice.

1. Preproof prep

When we first come face-to-face with a mathematical statement that we are trying to either prove or disprove, the first reaction is usually one of panic and possibly terror. All that reaction indicates is that we do not wholly understand the statement. Our main challenge is to understand mathematical statements at the deepest possible level. Remember: It is impossible to prove a statement that we do not understand! To understand a statement, we can consider special cases, look at specific examples, and recall the definitions of the words in the statement. Before we can prove a statement is true, we must believe it is true. Can we think of a heuristic argument or informal reasoning as to why we should believe it? If we can create a sketchy argument, then we might be well on our way to uncovering an ironclad proof.

If a statement is false, we need only produce a specific counterexample in which the hypotheses are satisfied while the conclusion does not hold. In the case of a false statement, our mission is to create a new statement that is true. We could imagine variations, include additional hypotheses, or weaken the conclusion to produce a provable statement.

One way to view creating a mathematical proof is as building a bridge. For us to get over the giant abyss of our ignorance, we must build a bridge from what we know to what we are trying to discover. Thus it is always critical to consider what we already know. In the context of this book, usually we can wonder if any of the previous challenges are useful in conquering the challenge at hand. In many cases, the previous challenge is the best hint to the current one.

The two key points to remember are: (1) we cannot prove a statement that we do not understand, and (2) we cannot start writing a mathematical proof until we have an idea of the logical thread that will be the core of the argument.

2. Penning the proof

Once we have an idea or approach to prove a statement, we must communicate our creative insights to others. We therefore need to write a polished narrative in prose. Almost always, the first attempt at writing a proof will not be the final version. Thus we should expect to start over again several times. In fact, we should consider our first attempt as just that—a first draft, understanding that it is not the final word.

The proof itself should consist of complete, grammatically correct sentences that guide the reader through our ideas and our logic. In reality, a proof is really a paper. There needs to be a beginning, a middle, and an end. The prose and ideas should be broken into paragraphs. Equations should be marked so they can be referred to in the argument, as needed. We should lay out our argument in a logical, flowing fashion. Again, we usually will not hit that right logical flow the first time out—it will evolve through several drafts. If you have seen shorthand symbols such as $\exists$ or $\forall$ or $\ni$ or $\therefore$, then congratulations, but do not use them in a final draft of a formal proof.

As we write a proof, we should always be sensitive to our audience—the reader. Are we writing for research mathematicians or undergraduate students or a high school math fan? Knowing our readership will allow us to craft our argument at the right level. I tell my students to write for other students in the class. Thus the arguments can be at a sophisticated level, but details (that might be omitted for a mathematician) should remain in the body of the proof. In addition, as the material increases in abstraction and difficulty, the sophistication of the proofs should probably increase as well.

In mathematical circles, the convention is to write in the first person plural. Thus we use "we" rather than "I." Also there are many standard words that should be imported into proofs when appropriate, because they help the logic flow. These words and phrases include "If…, then…"; "because…"; "we recall that…"; "and thus…"; "it follows that…"; "therefore…"; "hence…"; "whence…"; "let…"; "suppose…"; and "assume…". It is important to understand the meaning of each of these words and phrases so that they are used appropriately and correctly.

We should make it clear when we introduce a new parameter or variable. Thus by simply writing "$x + y = z$," it is not clear what, if anything, is being defined. It is better to be explicit: "Given integers x and y, let $z = x + y$." Above all, we should try to write as logically as possible. One idea should naturally lead to the next. The image we should keep in our minds is that we are carrying along a reader who does not know our ideas—that reader will come to understand our intuition and insight through our writing. It's our job to make that development as easy, painless, and interesting as possible.

3. Post proof

Once we have authored and edited a proof, the best course of action is to put it aside and reread it a day or so later. Thus it is key to start working on a proof well before it is due! When we reread it a day or so later, is it still clear? If not, then we need to further edit and revise. We shouldn't feel bad about this reality; it's part of the process. On good days, a second look will lead to an easier or more direct argument. We should always challenge ourselves to find new and improved arguments. Finally, we must face the following key question: Is the proof *correct, complete,* and *clear*? Once we are satisfied and proud (yes, proud) of our work, then we can call it complete and submit it for review.

Enjoy the creative challenge of discovering new ideas and finding your voice so that you can articulate your creativity for all to admire and appreciate.

A3

Commentary for instructors

In the opening of this book, I remarked that this discovery approach entails uncomfortable learning. That pedagogy also leads to uncomfortable teaching. By "uncomfortable," I mean difficult and sometimes awkward. As one might expect, crafting courses with such a heavy student-discovery component is extraordinarily challenging—especially for those instructors who have not experienced inquiry-based learning in their own education.

I would argue, however, that a discovery approach to mathematics is an extremely effective means for students to wrap their minds around deep ideas, make those ideas their own, and generate new ideas. And though I confess that I do not teach all my courses in a purely discovery-based manner, the experiences I have had using this style of inquiry-based learning have resulted in some of the best and most rewarding teaching experiences in my career. Here I offer some general observations regarding this pedagogical style. In the next section, I describe, in greater detail, my own experiences with using this book in the classroom.

There are many different ways to use this book with students. Obviously it could be used in any appropriate independent study work or as a supplement to any lecture-style course. In the latter case, modules would make nice projects to complement the course materials and offer a discovery-based component outside of class. Alternatively, instructors might wish to adopt a hybrid instructor-lecture/student-presentation format, in which the instructor provides lectures but also sets aside some class time for student presentations.

I will focus my remarks on a pure discovery approach in which there are essentially no lectures given by the instructor; obvious adaptations can be made for lecture-style or hybrid-style courses.

A standard reality with discovery-based approaches in the classroom is that less material is covered. In my experience, I have covered about 10 or 11 modules in our 13-week semester here at Williams College. It is important to keep the class moving forward but without cutting off presenters or dismissing comments from the class. It needs to be made clear to students that if they agree to make a presentation, then they are implicitly stating that they are prepared to deliver a clear, correct, and complete solution or proof. If students get bogged down at the chalkboard, they should know that they can always back down without losing face. The instructor should be encouraging but should quickly move on to another student or "table" the challenge for another day.

If comments and questions from the audience are allowed to get out of control, then they will! The audience should understand that their role is to carefully listen to the presentation and afterward answer: *Is the presented proof correct? Complete? Clear?* Questions from the audience should always be asked in a respectful and encouraging manner (rather than in a competitive or argumentative fashion). The classroom environment should be light, friendly, and ideally festive. A common comment from an audience member is, "I see how Kate proved the result, and it looks okay to me, but I did it another way. Can I show you what I did?" Of course, pursuing this direction will bring forward progress to a halt. I avoid this scenario by stating early on that we are focusing on the proof presented. If we all agree that *that* proof is correct, complete, and clear, then we sign off on it and move on. That is, the only proof we consider is the present proof. Other proofs are fine, and even great, and should be submitted when the written work is collected. If students wish to present their (alternative) arguments, invite them to do so in your office. The class must move on.

I believe this one-proof-per-theorem format also applies (sadly) to the instructor. If a student proves a result in a correct but convoluted manner or if there is a much more elegant argument that the instructor knows, it is tempting to show the "cool" or "correct" proof after the student presentation.

Such actions could undermine the students' confidence. Showing a slicker proof might be edifying, but I would argue it is also demoralizing. Therefore I avoid sharing my favorite proofs with the students (as difficult as that may be). Students should take pride in and ownership of their proofs. (I often name a result after the student who presented its proof. I've had, among many others, Son's Theorem, Nela's Theorem, and even the Matt Conjecture.)

My hope is that the challenges in this book lead to many interesting discussions between the instructor and the students. Beyond the ideas and mathematical vistas, conversations regarding issues of communicating mathematics are key. A central theme of my course is to allow students to find their own mathematical voice and produce well-written and clear arguments. Even explaining where and when we use words such as "since," "assume," "however," "because," "thus," "hence," "by," and "therefore" in the theorem-proving business is a most valuable endeavor.

In my own experience, some students often feel at the beginning of a semester as if they have been thrown into the deep end of the mathematics pool. I explicitly tell them that yes they have been thrown in, but I then assure them that their flailing to stay afloat will evolve into graceful strokes that will have them gliding through the charted and uncharted waters of mathematics. In the end, students rise up to the difficult challenge of this material, work extremely hard, feel very proud of their work, and truly appreciate their accomplishments.

How I used this material in my classroom

I developed most of this material for a one-semester course I created at Williams College in 2003. Below is the course description from the college catalog.

MATH 251: Introduction to Mathematical Proof and Argumentation

The fundamental focus of this course is for students to acquire the ability to create and clearly express mathematical arguments through an exploration of topics from discrete mathematics. Students will learn various mathematical proof techniques while discovering such areas as logic, number theory, infinity, geometry, graph theory, and probability.

> Our goal is not only to gain an understanding and appreciation of interesting and important areas of mathematics but also to develop and critically analyze original mathematical ideas and argumentation.
>
> *Prerequisites:* Calculus I or II or one year of high school calculus with permission of instructor.

The student population in the classes consisted primarily of first- and second-year students. Some students had taken linear algebra, and others had not. For most students, this was their first "proof course." I paired students into teams and determined the pairing via a Math Personality Questionnaire that I created (see Appendix 4). I also invited students to inform me privately if there were any students in the class they preferred not to be paired with or if they had a preference for their partner. Once students were paired up, their only allowable resources were their partners (with whom they should have met at least twice a week), this text, and me. All other resources and individuals were off limits.

Each student was responsible for individually writing up the solution to each challenge. I asked students to type their final solutions and proofs, because that encouraged them to proofread and edit their written submissions. Every challenge was discussed in class. If students submitted a solution on or before the class in which that solution was discussed, that solution was graded out of a possible 10 points. However, if the student submitted a solution after it was discussed in class (but no more than one week later), then it was graded out of 8 points. Thus there was a penalty for not completing the work before it was discussed in class, but that penalty was not fatal. Extensions to theorems did not require proofs, and students received bonus points for their generalizations.

I had an index card for each student, and I brought the stack of cards to every class. In class, I would select a student to show the group his or her solution or proof to the challenge at hand. Students could elect to pass if they had not completed the challenge, but I would note this on their index cards. For the student who offered a proof, I would note the date and the challenge and, after the presentation, note a grade from 1 to 10, depending on the quality of the student's argument, presentation, and ability to answer questions. I would select those students who had not presented recently—thus attempting to realize an even distribution of presentations among students.

The material I covered in my classes parallels that given in the opening section of this book, under the course Introduction to Mathematical Proof. Specifically, one semester, I had students explore (in this order) Modules 2, 15, 3, 4, 6, 7, 8, 9, 10, 11, 12, and a bit of 13. Another semester's syllabus was Modules 1, 2, 3, 4, 5 (first half), 6, 7, 8, 9, 10, and 11.

The final course grade was computed as follows: Class Presentations = 30%; Written Submissions = 20%; Midterm Exam = 25%; and Final Exam = 25%. The final examination was a 24-hour, open text, open notes exam. Both the midterm and the final had a computational component, as well as a theoretical (prove theorems) section.

I hope this commentary provides a starting point for those instructors who are considering offering their students this material in a hands-on approach. More important, I hope that it inspires instructors to craft the perfect courses for themselves and their students and that everyone enjoys the challenges and triumphs involved in discovering the wonderful world of mathematical ideas.

A4

A math personality questionnaire

Those readers who wish to collaborate with another person or groups of people might find it useful to answer the following questionnaire. It could help an instructor or group leader find individuals with compatible mathematical personalities, potentially leading to happy partnerships.

For each statement, please assign a number response from 0 to 10: 0 = disagree strongly; 5 = maybe/maybe not; 10 = agree strongly. **There are no correct answers!** Answer as honestly as you can.

1. I'm an outgoing person.

2. I am extremely competitive.

3. I'm a bit insecure of my mathematical abilities.

4. I hope my partner will not ask a bunch of silly questions.

5. Math is not my number one priority.

6. I like thinking about math by myself first before talking with others.

7. I sometimes feel I work slower than others.

8. I like chatting with friends about math homework and then thinking about the questions on my own.

9. I prefer to work on my own.

10. I like learning from others.

11. A math partnership should be a give-and-take relationship.

12. I wish I could do math all the time.

13. I want to consider my partner as a resource.

14. I hope I'll be able to interact with my partner on a regular basis.

15. I love being right.

16. I love learning something new from a classmate.

17. I love making mistakes because I almost always gain some new insight.

18. I am really excited about thinking about mathematics.

19. I'd like my partner to be low-key.

20. I like working on math after midnight.

21. I like working on math at the last minute.

22. Once I find a solution, I like to think more deeply about the issue.

23. I plan to study more math after this course.

24. I like to start on my homework as soon as possible.

25. I think my mathematics professor is the cat's pajamas.

Acknowledgments

My three classes of Math 251: Introduction to Mathematical Proof and Argumentation provided me with some of my most joyful classroom experiences. There was so much laughter and high spirits in class that instructors from adjacent classrooms would often inquire as to what was going on in my course. The students from these classes were a constant source of inspiration and encouragement as they worked through various manuscript versions of this book. Their comments, suggestions, and reactions greatly improved this text. Thus I am delighted to acknowledge and thank these wonderful students, who not only made this book a reality but who also helped make its creation so pleasurable.

From my fall 2003 Math 251 class, I thank Andrea Burke, Dan Burns, Emily Button, Alexandra Constantin, Neal Holtschulte, Thomas Kunjappu, Ruoweng Liu, Karl Naden, Aaron Pinsky, Paul Stansifer, and Bartley Tablante.

From my spring 2005 Math 251 classes, I thank Bill Bernsen, Mack Brickley, Jing Cao, Nathan Cook, Matthew Earle, Isaac Gerber, Michael Gnozzio, Paul Hess, Son Ho, Michael Kamida, Jesse Levitt, Sean McKenzie, Katherine Nolfi, Gordon Phillips, Will Sheridan, Kim Taylor, Nela Vukmirovic, Andy Whinery, and Benjamin Wood.

I wish to thank all those who were generous enough to review various drafts of this text and offer me valuable feedback. They include Daniel King, Lew

Ludwig, Frank Morgan, Cornelius Stallmann, Rob Tubbs, and Eric Stade. Edward Dunne, Peter Jipsen, and Barry Spieler carefully studied the entire manuscript and offered many important suggestions. Allison Pacelli taught from a manuscript version of this text, provided a number of excellent suggestions, and was the inspiration for the "Proof Primer" in Appendix 2. I wish to thank my students Matthew Earle and Aaron Pinsky for their assistance in outlining solutions to all the challenges. I especially wish to acknowledge and thank my colleague Deborah Bergstrand, for her tireless efforts in assisting me with the instructor resources that appear as a separate publication and for the almost uncountably many helpful suggestions.

Finally, I thank my copyeditor, Tara Joffe, for her valuable contributions. From Key College Publishing, I express my gratitude to Jensen Barnes, Richard Bonacci, Kristin Burke, Allyndreth Cassidy, Laura Ryan, and Mike Simpson for their creative contributions and their enthusiasm for this project from the very beginning.

Financial support for this project was provided, in part, by a grant from Williams College and by a grant from the Educational Advancement Foundation. I wish to acknowledge Mr. Harry Lucas, Jr., from the Educational Advancement Foundation, for his constant support, enthusiasm, and encouragement. For well over 30 years, Mr. Lucas has actively championed the cause of inquiry-based learning, and his creative efforts continue to bring real and significant change to the world of mathematics education. Thus it is a privilege for me to thank Mr. Harry Lucas, Jr., who has been a constant source of inspiration.

Index

D

E

F

G

H

I

J

K

L

M

N

O

P

Q

R

T

U

V

W